Mineral Fibres

Mineral Fibres

Special Issue Editors

Andrea Bloise
Rosalda Punturo
Robert Kusiorowski
Lola Pereira

MDPI • Basel • Beijing • Wuhan • Barcelona • Belgrade

MDPI

Special Issue Editors

Andrea Bloise
Department of Biology, Ecology and
Earth Sciences, University of Calabria,
Italy

Rosalda Punturo
Departmen Biology, Geology, Natural
Environment, University of Catania,
Italy

Robert Kusiorowski
Institute of Ceramics and Building
Materials Refractory Materials Division,
Poland

Lola Pereira
Departmentt of Geology, University
of Salamanca,
Spain

Editorial Office
MDPI
St. Alban-Anlage 66
4052 Basel, Switzerland

This is a reprint of articles from the Special Issue published online in the open access journal *Fibers* (ISSN 2079-6439) in 2019 (available at: http://www.mdpi.com/si/fibers/Mineral_Fibres).

For citation purposes, cite each article independently as indicated on the article page online and as indicated below:

LastName, A.A.; LastName, B.B.; LastName, C.C. Article Title. *Journal Name* **Year**, *Article Number*, Page Range.

ISBN 978-3-03921-144-9 (Pbk)
ISBN 978-3-03921-145-6 (PDF)

Contents

About the Special Issue Editors . . . . **vii**

Andrea Bloise, Rosalda Punturo, Robert Kusiorowski and Dolores Pereira Gómez
Editorial for Special Issue "Mineral Fibres"
Reprinted from: *Fibers* **2019**, *7*, 54, doi:10.3390/fib7060054 . . . . **1**

Maria Carmela Dichicco, Michele Paternoster, Giovanna Rizzo and Rosa Sinisi
Mineralogical Asbestos Assessment in the Southern Apennines (Italy): A Review
Reprinted from: *Fibers* **2019**, *7*, 24, doi:10.3390/fib7030024 . . . . **4**

Jerzy Witek, Bronisław Psiuk, Zdzisław Naziemiec and Robert Kusiorowski
Obtaining an Artificial Aggregate from Cement-Asbestos Waste by the Melting Technique in an
Arc-Resistance Furnace
Reprinted from: *Fibers* **2019**, *7*, 10, doi:10.3390/fib7020010 . . . . **17**

Rosalda Punturo, Claudia Ricchiuti and Andrea Bloise
Assessment of Serpentine Group Minerals in Soils: A Case Study from the Village of San
Severino Lucano (Basilicata, Southern Italy)
Reprinted from: *Fibers* **2019**, *7*, 18, doi:10.3390/fib7020018 . . . . **31**

Rosalda Punturo, Claudia Ricchiuti, Marzia Rizzo and Elena Marrocchino
Mineralogical and Microstructural Features of Namibia Marbles: Insights about Tremolite
Related to Natural Asbestos Occurrences
Reprinted from: *Fibers* **2019**, *7*, 31, doi:10.3390/fib7040031 . . . . **43**

**Gaia Maria Militello, Andrea Bloise, Laura Gaggero, Gabriele Lanzafame and
Rosalda Punturo**
Multi-Analytical Approach for Asbestos Minerals and Their Non-Asbestiform Analogues:
Inferences from Host Rock Textural Constraints
Reprinted from: *Fibers* **2019**, *7*, 42, doi:10.3390/fib7050042 . . . . **56**

**Andrea Bloise, Claudia Ricchiuti, Eugenia Giorno, Ilaria Fuoco, Patrizia Zumpano,
Domenico Miriello, Carmine Apollaro, Alessandra Crispini, Rosanna De Rosa and
Rosalda Punturo**
Assessment of Naturally Occurring Asbestos in the Area of Episcopia (Lucania, Southern Italy)
Reprinted from: *Fibers* **2019**, *7*, 45, doi:10.3390/fib7050045 . . . . **72**

**Miguel A. Rivero Crespo, Dolores Pereira Gómez, María V. Villa García,
José M. Gallardo Amores and Vicente Sánchez Escribano**
Characterization of Serpentines from Different Regions by Transmission Electron Microscopy,
X-ray Diffraction, BET Specific Surface Area and Vibrational and Electronic Spectroscopy
Reprinted from: *Fibers* **2019**, *7*, 47, doi:10.3390/fib7050047 . . . . **83**

Oliviero Baietto, Mariangela Diano, Giovanna Zanetti and Paola Marini
Grinding Test on Tremolite with Fibrous and Prismatic Habit
Reprinted from: *Fibers* **2019**, *7*, 52, doi:10.3390/fib7060052 . . . . **94**

About the Special Issue Editors

Andrea Bloise is currently a researcher in mineralogy at the Department of Biology, Ecology, and Earth Sciences (DiBEST) at Calabria University, Rende, Italy. Since 2002, his research activity has embraced the following topics: (i) hydrothermal synthesis of doped phases with a special attention to sheet silicates, asbestos minerals, borate, and titanosilicate; (ii) flux growth at high temperature of dope and pure silicate with the aim to test their physical–chemical and technological properties; and (iii) asbestos and other fibrous natural and synthetic minerals characterization. He has expertise in X-ray powder diffraction (XRPD), scanning and transmission electron microscopy (SEM and TEM), and thermal analysis (TG/DSC) collaborating on various scientific projects of national and international interest. He has recently also been working in the fields of cultural heritage and geochemical modeling of both natural and thermal waters. He is the author of several international peer-reviewed scientific publications and scientific director of the Laboratory of Experimental Mineralogy at the DiBEST.

Rosalda Punturo is a researcher in Petrology and Petrography and Assistant Professor at the University of Catania, Italy, responsible for the Laboratory of Geochemistry at the Department of Biological, Geological and Environmental Sciences. She has a Ph.D in Igneous Petrology (2000) on petrophysical and petrographic properties of deep seated xenoliths and is the national coordinator of the Research Project of National Interest of the Italian Ministry of University entitled: "Strain rate in mylonitic rocks and induced changes in petrophysical properties across the shear zones". Her scientific achievements include over 50 publications in different indexed journals and various presentations at international and national scientific conferences. She has also organized and served as the convenor of thematic sessions at national and international congresses. The current research interests include asbestos and asbestiform minerals, petrophysical properties of minerals and rocks, and environmental issues.

Robert Kusiorowski earned his Ph.D. degree in chemical technology at Silesian University of Technology in 2014, where he worked on the thermal decomposition process and disposal of cement-asbestos wastes. He is currently an Assistant Professor in Łukasiewicz Research Network, Institute of Ceramics and Building Materials, Refractory Materials Division in Gliwice, Poland. His scientific achievements include over 20 publications in different journals, two chapters of the books, and over 20 presentations at international and national scientific conferences. The current research interests include: ceramic building materials, asbestos issues, refractories, thermal insulation materials, binders, and recycling technologies.

Dolores Pereira is professor of Geology and Engineering Geology at the University of Salamanca. She also teaches at the Master of Social Studies for Science and Technology of this university. She is Secretary General of the IUGS Heritage Stones Subcommission, Leader of the IGCP-637, Chair of the IUGS Publications Committee, and Member of the Books Editorial Committee of the Geological Society of London. Her research interests include mineralogy, petrology, characterization of natural stones for construction and restoration and their importance on architectural heritage, and also bibliometrics and citation analysis. Dolores Pereira is vice-president of AMIT, the Spanish Association for Women in Science.

fibers

MDPI

Editorial

Editorial for Special Issue "Mineral Fibres"

Andrea Bloise [1],*, Rosalda Punturo [2], Robert Kusiorowski [3] and Dolores Pereira Gómez [4]

[1]   Department of Biology, Ecology and Earth Sciences, University of Calabria, Via Pietro Bucci,
      I-87036 Rende, Italy
[2]   Department Biology, Geology, Natural Environment, University of Catania, Corso Italia, 55,
      95129 Catania, Italy; punturo@unict.it
[3]   Łukasiewicz Research Network, Institute of Ceramics and Building Materials Refractory Materials Division,
      Toszecka 99, 44-100 Gliwice, Poland; r.kusiorowski@icimb.pl
[4]   Department of Geology, University of Salamanca, Pl. De la Merced s/n, 37008 Salamanca, Spain; mdp@usal.es
*   Correspondence: andrea.bloise@unical.it; Tel.: +39-0984-493588

Received: 9 June 2019; Accepted: 11 June 2019; Published: 13 June 2019

In the past 30 years, there has been a growing concern regarding the health risks of exposure to asbestos-containing materials (ACMs) and naturally occurring asbestos (NOA). Nowadays, harmful asbestos minerals that are regulated by law (in Europe and in several countries worldwide) include fibrous forms of the minerals chrysotile, crocidolite, amosite, tremolite, actinolite and anthophyllite. Asbestos has been classified as a Group 1 carcinogenic material by the International Agency for Research on Cancer. Therefore, many countries have banned its production and use [1]. In the past, more than 3000 types of asbestos-containing materials (ACMs) were used for making a wide variety of products for cinemas, schools, hospitals and army equipment, as well as in many industrial applications [1], because of their thermal insulation properties. However, the mining and use of tremolite asbestos, actinolite asbestos and anthophyllite asbestos was reduced compared to more commercially available types of asbestos such as chrysotile, crocidolite and amosite. Commercial asbestos fibres remain a serious problem both in previous installation in manufactured goods and in those countries where asbestos is still used. In the countries that banned the use of asbestos minerals and where remediation policies are encouraged, many studies and patents have dealt with the possible disposal and re-use of ACMs. The proposed inertization/recycling methods include thermal treatment, mechanical treatments, chemical, biological and biochemical treatments [2,3]. The preference for recycling compared to landfill disposal is specified in the European Directive [4], since recycling is the best solution, as it reduces environmental impact and the consumption of primary raw materials.

Although cases of disease due to exposure to ACMs may be decreasing in many countries of the world, there are newly recognized sources that pose a serious public health problem. These are unanimously called "naturally occurring asbestos" (NOA) [5], a term used to describe natural sources that trigger risk for a population due to weathering or human activities that produce dust consisting of fibrous minerals that may or may not fit the regulatory definitions of asbestos. In this regard, it is worth mentioning that non-regulated fibres (asbestiform) such as erionite, ferrierite and fluoro-edenite, are considered to be positive carcinogenic minerals sometimes more dangerous than the six regulated asbestos fibres [6–8]. Indeed, the US National Institute for Occupational Safety and Health (NIOSH) has recently proposed to extend the definition of asbestos to all of the elongated mineral particles (EMP). In many geological formations and outcrops, asbestos and EMP usually occur together [9] and these minerals must be discriminated correctly from a morphological point of view. In this regard, NIOSH highlights the difficulty in ascertaining the source of exposure in the case of mixed exposures for some mining operations.

NOA detection and quantification actions are important for providing the administrative agencies useful knowledge in order to carry out protocols for exposure control during construction such as highways, civil constructions and retaining walls. Many communities worldwide are potentially exposed to NOA, e.g., [10–18], and consciousness is increasing. In fact, health effects caused by NOA exposure continue to be of great public interest since the increasing risk of health problems for people who live close to NOA deposits worldwide has been widely demonstrated. In the last few years, excessive incidences of lung cancer and malignant mesothelioma have been reported as a consequence of the presence of NOA in Italy, Turkey, Greece, Corsica, New Caledonia, USA and China [19]. A crucial theme of interest related to environmental pollution is the enhanced mobilization of asbestos or asbestiform minerals affecting soils and rocks, due to human activities (e.g., road construction, excavation, mining) in comparison with natural weathering processes. Moreover, when natural causes or anthropic factors affect rocks which host asbestos, some naturally occurring harmful elements (e.g., Cr, Ni, Co, V) may be disseminated in the environment, resulting in the contamination of soil, water and air.

In summary, this Special Issue entitled "Mineral Fibers" depicts the state of the art about NOA as a source of possible environmental risks for populations, due to the adverse health effects associated with exposure. Case studies from various geological contexts are presented together with contributions presenting novel and classical approaches for asbestos inertization and recycling, together with possible solutions for reducing asbestos exposure.

Conflicts of Interest: The authors declare no conflict of interest.

References

1. Gualtieri, A.F. *Mineral Fibers: Crystalchemistry, Chemical-Physical Properties, Biological Interaction and Toxicity*; European Mineralogical Union and Mineralogical Society of Great Britain and Ireland: London, UK, 2017; p. 533.
2. Spasiano, D.; Pirozzi, F. Treatments of asbestos containing wastes. *J. Environ. Manag.* **2017**, *204*, 82–91. [CrossRef] [PubMed]
3. Bloise, A.; Kusiorowski, R.; Lassinantti Gualtieri, M.; Gualtieri, A.F. Thermal behaviour of mineral fibers. In *Mineral Fibers: Crystal Chemistry, Chemical-Physical Properties, Biological Interaction and Toxicity*; Gualtieri, A.F., Ed.; European Mineralogical Union: London, UK, 2017; Volume 18, pp. 215–252.
4. The European Parliament and the Council of the European Union. Directive 2008/98/EC of the European Parliament and of the Council of 19 November 2008 on waste and repealing certain Directives (Text with EEA relevance). *Off. J. Eur. Union* **2008**, *L312*, 3–30.
5. Harper, M. 10th Anniversary critical review: Naturally occurring asbestos. *J. Environ. Monit.* **2008**, *10*, 1394–1408. [CrossRef] [PubMed]
6. Baumann, F.; Ambrosi, J.-P.; Carbone, M. Asbestos is not just asbestos: An unrecognised health hazard. *Lancet Oncol.* **2013**, *14*, 576–578. [CrossRef]
7. Ballirano, P.; Bloise, A.; Gualtieri, A.F.; Lezzerini, M.; Pacella, A.; Perchiazzi, N.; Dogan, M.; Dogan, A.U. The crystal structure of mineral fibers. In *Mineral Fibers: Crystal Chemistry, Chemical-Physical Properties, Biological Interaction and Toxicity*; Gualtieri, A.F., Ed.; European Mineralogical Union: London, UK, 2017; Volume 18, pp. 17–53.
8. Gualtieri, A.F.; Gandolfi, N.B.; Passaglia, E.; Pollastri, S.; Mattioli, M.; Giordani, M.; Ottaviani, M.F.; Cangiotti, M.; Bloise, A.; Barca, D.; et al. Is fibrous ferrierite a potential health hazard? Characterization and comparison with fibrous erionite. *Am. Miner.* **2018**, *103*, 1044–1055. [CrossRef]
9. Belluso, E.; Cavallo, A.; Halterman, D. Crystal habit of mineral fibres. In *Mineral Fibres: Crystal Chemistry, Chemical-Physical Properties, Biological Interaction and Toxicity*; Gualtieri, A.F., Ed.; European Mineralogical Union: London, UK, 2017; Volume 18, pp. 65–109.
10. Bloise, A.; Punturo, R.; Catalano, M.; Miriello, D.; Cirrincione, R. Naturally occurring asbestos (NOA) in rock and soil and relation with human activities: The monitoring example of selected sites in Calabria (southern Italy). *Ital. J. Geosci.* **2016**, *135*, 268–279. [CrossRef]
11. Bloise, A.; Belluso, E.; Critelli, T.; Catalano, M.; Apollaro, C.; Miriello, D.; Barrese, E. Amphibole asbestos and other fibrous minerals in the meta-basalt of the Gimigliano-Mount Reventino Unit (Calabria, south-Italy). *Rend. Online Soc. Geol. It.* **2012**, *21*, 847–848.

12. Petriglieri, J.R.; Laporte-Magoni, C.; Gunkel-Grillon, P.; Tribaudino, M.; Bersani, D.; Sala, O.; Salvioli-Mariani, E. Mineral fibres and environmental monitoring: A comparison of different analytical strategies in New Caledonia. *Geosci. Front.* **2019**, in press. [CrossRef]

13. Dichicco, M.C.; Paternoster, M.; Rizzo, G.; Sinisi, R. Mineralogical asbestos assessment in the Southern Apennines (Italy): A Review. *Fibers* **2019**, *7*, 24. [CrossRef]

14. Bloise, A.; Catalano, M.; Critelli, T.; Apollaro, C.; Miriello, D. Naturally occurring asbestos: Potential for human exposure, San Severino Lucano (Basilicata, Southern Italy). *Environ. Earth Sci.* **2017**, *76*, 648. [CrossRef]

15. Bloise, A.; Ricchiuti, C.; Giorno, E.; Fuoco, I.; Zumpano, P.; Miriello, D.; Apollaro, C.; Crispini, A.; De Rosa, R.; Punturo, R. Assessment of naturally occurring asbestos in the area of Episcopia (Lucania, Southern Italy). *Fibers* **2019**, *7*, 45. [CrossRef]

16. Buck, B.J.; Goossens, D.; Metcalf, R.V.; McLaurin, B.; Ren, M.; Freudenberger, F. Naturally occurring asbestos: Potential for human exposure, Southern Nevada, USA. *Soil Sci. Soc. Am. J.* **2013**, *77*, 2192–2204. [CrossRef]

17. Punturo, R.; Bloise, A.; Critelli, T.; Catalano, M.; Fazio, E.; Apollaro, C. Environmental implications related to natural asbestos occurrences in the ophiolites of the Gimigliano-Mount Reventino Unit (Calabria, southern Italy). *Int. J. Environ. Res.* **2015**, *9*, 405–418.

18. Rivero Crespo, M.A.; Pereira Gómez, D.; Villa García, M.V.; Gallardo Amores, J.M.; Sánchez Escribano, V. Characterization of serpentines from different regions by transmission electron microscopy, X-ray diffraction, BET specific surface area and vibrational and electronic spectroscopy. *Fibers* **2019**, *7*, 47. [CrossRef]

19. Case, B.W.; Marinaccio, A. Epidemiological approaches to health effects of mineral fibres: Development of knowledge and current practice. In *Mineral Fibers: Crystal Chemistry, Chemical-Physical Properties, Biological Interaction and Toxicity*; Gualtier, A.F., Ed.; European Mineralogical Union: London, UK, 2017; Volume 18, pp. 376–406.

fibers

MDPI

Review

Mineralogical Asbestos Assessment in the Southern Apennines (Italy): A Review

Maria Carmela Dichicco [1], Michele Paternoster [1,2], Giovanna Rizzo [1,*] and Rosa Sinisi [1]

[1] Department of Sciences, University of Basilicata, Campus di Macchia Romana, Viale dell'Ateneo Lucano 10, 85100 Potenza, Italy; maria.dichicco@unibas.it (M.C.D.); michele.paternoster@unibas.it (M.P.); rosa.sinisi@unibas.it (R.S.)

[2] Istituto Nazionale di Geofisica e Vulcanologia, Sezione di Palermo, Via Ugo La Malfa 153, 90146 Palermo, Italy

* Correspondence: giovanna.rizzo@unibas.it; Tel.: +39-0971-20-5833

Received: 14 February 2019; Accepted: 13 March 2019; Published: 19 March 2019

Abstract: This paper deals with petrography and mineralogy of serpentinitic rocks occurring in the Southern Apennines (Italy) with the aim to review the already available literature data and furnish new details on asbestos minerals present in the studied area. Two sites of Southern Italy were taken into account: the Pollino Massif, at the Calabrian-Lucanian border, and the surroundings of the Gimigliano and Mt. Reventino areas where serpentinites of Frido Unit are mainly exposed. Textural and mineralogical features of the studied rocks point to a similar composition for both sites including asbestos minerals such as chrysotile and tremolite-actinolite series mineral phases. Only in the Pollino Massif serpentinites edenite crystals have been detected as well; they are documented here for the first time. This amphibole forms as fibrous and/or prismatic crystals in aggregates associated with serpentine, pyroxene, and calcite. Metamorphism and/or metasomatic alteration of serpentinites are the most probable processes promoting the edenite formation in the Southern Apennine ophiolitic rocks.

Keywords: asbestos' minerals; edenite; serpentinites; Southern Italy

1. Introduction

In the last decade, many researchers have focused on serpentinites cropping out in the ophiolitic sequences and have aimed to assess and monitor their potential as asbestos-bearing lithotypes, since asbestos occurrence in mafic and ultramafic rocks that undergo ocean floor metamorphism is relatively common [1–4]. In Italy, the occurrence of these rocks is documented both in the Alps and Apennines. These outcrops extend from the Ligurian-Piedmont through the Tuscan-Emilian Apennines as far as Val Tiberina and continue, in disjointed groupings, to the Calabrian-Lucanian Apennines [5].

As is known, the definition of asbestos used by regulatory agencies [6] for identification includes the following six mineral species: chrysotile, crocidolite, tremolite, actinolite, amosite, and anthophyllite [7]. Among these minerals, only chrysotile is a sheet silicate; the other minerals are included within the amphibole supergroup. Silicate minerals belonging to the serpentine and amphibole groups are flexible, heat-resistant, and chemically inert. These minerals usually occur with an elongated and/or bladed prismatic habit, although they may be acicular or fibrous as well. In the European countries, fibers having a length ≥ 5 μm, a width <3 μm, and an aspect ratio >3 are defined as "asbestos" by Directive 2003/18/CE.

Asbestos is classified as a carcinogen material of Category 1 by the world health authorities [8]. Several authors ascribe the fibers' toxicity to their morphology and size, chemical-physical characteristics, surface reactivity, and biopersistence [9]. It is known that the presence of impurities (i.e., Fe, Ni, and Ti) in the ideal chemical composition in asbestos fibers, even in small amounts, affects their chemical and physical properties, size, and shape [10–13]. Moreover, according to in vitro studies on biological-system–mineral interactions, both characteristics (impurities and size) are considered to be responsible for its pathological effects [14,15].

In this paper, we present data related to petrography and mineralogy of serpentinites in representative sites at the Pollino Massif (Calabrian-Lucanian boundary) and Gimigliano-Mt. Reventino (Sila Piccola, Northern Calabria), with the principal aim to review the main modes of occurrence of asbestos minerals in the Southern Apennines.

2. Ophiolitic Sequences in Southern Italy

The Southern Apennines is a fold-and-thrust chain developed between the Upper Oligocene and the Quaternary during the convergence between the African and European plates [16–18]. The ophiolitic sequences incorporated in the Southern Apennine chain are related to the northwest subduction of the oceanic lithosphere pertaining to the Ligurian sector (divided in the Frido Unit and North Calabrian Unit) of the Jurassic Western Tethys. They crop out in the northeastern slope of the Pollino Ridge, along with the Calabria-Lucanian border zone, and in the Gimigliano-Mt. Reventino Unit (Sila Piccola, Southern Italy) (Figure 2). At the Calabrian-Lucanian boundary, the investigated sites are on well-exposed outcrops along road cuts, active and inactive quarries or in the proximity of villages (Pietrapica quarry, Timpa Castello quarry, Fagosa quarry, and Fosso Arcangelo, San Severino Lucano, Rovine Convento Sagittale localities, Mt. Nandiniello and Ghiaia quarry) (Figure 1), whereas close to the Gimigliano town the outcrops are in correspondence of quarries at Sila Piccola, Northern Calabria [5,19–21].

Figure 1. Serpentinite outcrops of the Pollino Massif: (**a**) Fagosa quarry, (**b**) Mt. Nandiniello, (**c**) Fosso Arcangelo site, and (**d**) San Severino Lucano site.

In the following sections, details on geological setting and formations of both sites are presented.

2.1. The Pollino Massif Serpentinites

According to several authors [22–25], the Frido Unit forms the uppermost thrust sheet of southern Apennines and tectonically overlies the North Calabrian Units, which in turn are split in different thrust sheets [25]. The Frido Unit is characterized by HP/LT metamorphic sequences developed between Upper Jurassic to the Upper Oligocene [26–29] and references therein. The ophiolitic rocks in the Frido Unit, from the bottom to the top, consist of tectonized serpentinite [30–35], metabasalt [36], metagabbro, metapillow lavas [37], dismembered metadoleritic and rodingite dykes [28,38–40], and sedimentary cover [41]. The serpentinites are englobed in tectonic slices and are associated with metadolerite dykes and continental crust rocks that mainly consist of weathered granofels, garnet gneiss, garnet–biotite gneiss, leucocratic biotite gneiss, and lenticular bodies of amphibolite [42]. As suggested by Knott [23], the Frido Unit underwent a polyphase blueschist to greenschist facies metamorphism developed in the deeper portions of the Liguride accretionary wedge.

In the Pollino Massif, the serpentinite rocks are cataclastic and massive. Cataclastic serpentinites show a high degree of fracturing and deformation. The millimeter to centimeter fractures are almost filled by exposed white and grey fibrous minerals [30,32,33]. Fibers occur as both large and elongate minerals developed over slickensided surfaces and/or as very fine-grained phases pervading the whole rock. Massive serpentinites show a low fracturing and deformation without exposed fibers.

2.2. The Gimigliano-Mt. Reventino Serpentinites

The Gimigliano-Mt. Reventino (Sila Piccola, Figure 2) occurs in the northern sector of the Calabrian-Peloritan Orogen [43,44]. According to Ogniben [45,46], the Northern Calabria sector consists of three main tectonic complexes: the Apennine Units Complex, at the bottom, made up of Mesozoic sedimentary and metasedimentary terranes; the allochthonous Alpine Liguride Complex, in the intermediate position, consisting of a series of Cretaceous-Paleogene metamorphic units that include metapelites, ophiolites, and carbonates; the Calabride Complex, at the top, with granites, gneisses, and metasedimentary deposits derived from Hercynian and pre-Hercynian terranes. The Mt. Reventino area is characterized mainly by lenses of metabasalts and serpentinites limited by low angle tectonic systems, with metapelites, metalimestones, and metarenites of uncertain ages of the Frido Unit (Liguride Complex). The massive-banded metabasalts and serpentinites lenses constitute the upper part of Mt. Reventino [43]. In the ophiolitic bodies, ascribed to the Liguride Complex of oceanic derivation [46–49], the serpentinites occupy the cores of the major tight folds and are partially or completely surrounded by isolated bodies of metabasalts and subordinate metadolerites [43]. In the Gimigliano-Mt. Reventino two different types of serpentinites occur as foliated and massive rocks. Mostly dark green serpentinites crop out as massive bodies that only sometimes are weakly foliated and cut by serpentine and calcite veins [5,19–21].

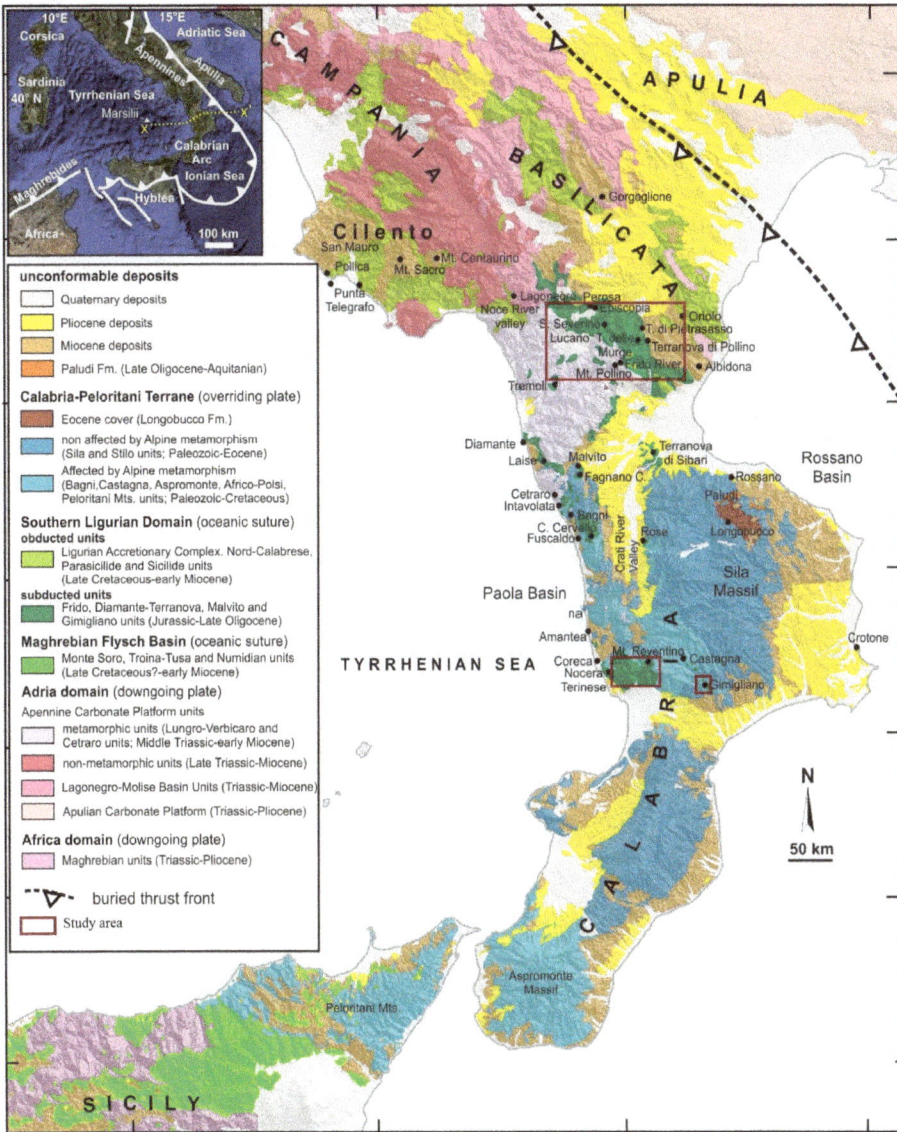

Figure 2. Geological sketch map of the Southern Apennines-Calabria-Peloritani chain and location of the study areas (modified after [50]).

3. Analytical Methods

In this paper, we report and discuss data available in literature referring to the petrographical and mineralogical studies performed on serpentinites from selected sites of southern Apennines. In particular, data here presented are from Dichicco et al. [32,33], Punturo et al. [19], Bloise et al. [5], and Campopiano et al. [20].

The petrographic characterization was carried out by optical microscopy on thin sections of rock samples. Percentages for fibrous minerals have been calculated by means of point-counting modal analysis following the EPA/600/R93-116 method. The mineralogical compositions have been obtained by using X-ray diffraction (XRD) on bulk rock powder. Specific analyses on single minerals were performed by μ-Raman spectroscopy, FT-IR spectroscopy, SEM-EDS, and electron microprobe (EMP) analyses. Details of analytical conditions are reported in the following papers: Dichicco et al. [32,33], Punturo et al. [19], and Bloise et al. [5].

4. Previous Studies and New Findings

4.1. Asbestos Minerals in Serpentinites from the Pollino Massif

Serpentinites are characterized by an original pseudomorphic texture and mylonitic-cataclastic structures (Table 1). They are made up of fibrous minerals accounting for the 55% of the total mineral composition. The mineralogical assemblage consists of serpentine group minerals, amphibole minerals (mainly tremolite-actinolite series), titanite, clinopyroxene, clinochlore, magnetite, Cr-spinel, talc, quartz, and carbonate phases. The serpentinites are cross-cut by a micro-network of nanometer to millimeter veins filled by fibrous serpentine and serpentine ± amphiboles, amphibole minerals, and calcite ± amphiboles [32]. Chrysotile occurs as short and fine-fibers in the matrix and in the contact between vein and rock. Chrysotile occurs preferentially in serpentinites that have undergone some degree of recrystallization, in which the serpentine minerals have developed interlocking microstructures. Primary magmatic clinopyroxene occurs in partially preserved grains. The amphibole shows acicular, fibrous, and elongated habitus and forms in veins and/or in the rock matrix as crowns around the clinopyroxene porphyroclasts [32].

Table 1. Textural features and mineralogical assemblages of serpentinites from the study areas. Abbreviations of mineral names are from Whitney and Evans [51].

Locality	Texture	Mineral Assemblage	Fibrous Minerals
Pollino Massif [1]	Pseudomorphic texture and mylonitic-cataclastic structures	Srp ± Mag ± Tr-Act- Ed *-Hbl ± Clc ± Cpx ± Spl ± Ttn ± Cal ± Dol ± Tlc ± Qz	Tremolite, antigorite, chrysotile, edenite*
Gimigliano-Mt. Reventino [2]	Protogranular texture	Srp ± Mag ± Tr-Act ± Chl ± Cpx ± Spl ± Cal ± Tlc	Tremolite, antigorite, chrysotile

[1] Data from [32,33,52], [2] Data from [5,19–21], * This study.

Serpentine (lizardite, chrysotile, and antigorite) and amphibole-like (actinolite, d = 8.31 Å; tremolite, d = 2.94 Å) minerals have been detected by XRD analysis and represent the dominant phases of the studied samples. The 2:1 phyllosilicate (clinochlore, d = 4.74 Å) and iron oxides (magnetite, d = 2.52 Å) also occur as subordinate phases along with different types of carbonates (calcite, d = 3.04; aragonite, d = 3.38; dolomite, d= 2.88) (Table 2).

Table 2. Mineral assemblage of the studied serpentinites as detected by XRD, where (+ + +) = major phase, (++) = minor phase (<10%), (+) = trace phase, and (−) = absent.

Locality	Serpentine	Magnetite	Amphibole	Carbonate	Pyroxene	Talc	Quartz	Titanite	Spinel	Clinochlore
Pollino Massif [1]	+++	++	++	++	+	+	+	+	++	++
Gimigliano-Mt. Reventino [2]	+++	+	++	+	+	+	−	−	+	++

[1] This study, [2] Data from [5,19–21].

Serpentine group minerals were also identified by μ-Raman spectroscopy. Chrysotile is distinguished from the other minerals of the serpentine group by means of an antisymmetric band at about 3699 cm^{-1}, with a tail toward lower wavenumbers, and a less pronounced peak at about 3691 cm^{-1} [32]. As reported by Dichicco et al. [32,33], different types of amphibole minerals also occur in the analyzed rocks. In the μ-Raman spectra of the OH vibrational region, the amphibole shows two peaks, the most intense of which is at 3675–3673 cm^{-1} (Mg; Mg; Mg), the second most intense at 3660–3663 cm^{-1} (Mg; Mg; Fe). The number and relative intensity of these bands represent pure tremolite and almost pure tremolite with a small percentage of Fe^{2+} (Fe-tremolite). The presence of Fe^{2+} is confirmed by FT-IR [33]. No Fe^{3+} is present, owing to the absence of absorption bands at $\Delta = 50$ cm^{-1} from the tremolite reference band in the FT-IR spectrum [33,52–56].

Secondary Electron observations by ESEM analyses show asbestos tremolite fibers that are straight, flexible and approximately 100 μm in length. The EDS chemical analysis shows that amphibole crystals are homogeneous, without zoning, although some crystals display different amounts of SiO$_2$, CaO, MgO, Fe$_2$O$_3$, Al$_2$O$_3$, and Na$_2$O in the rim and core [33].

The microchemical composition of most amphiboles detected by EMPA is typical of Ca-amphiboles, including tremolite and Mg-Fe-hornblende (Table 3) [56].

In addition, the EMP analysis revealed for the first time the presence of edenite in the serpentinites rocks of the Frido Unit. As shown in Figure 3, in the serpentinites of the Pollino Massif, edenite crystals grow with a fibrous habitus and form aggregates often associated with serpentine, diopside, and calcite.

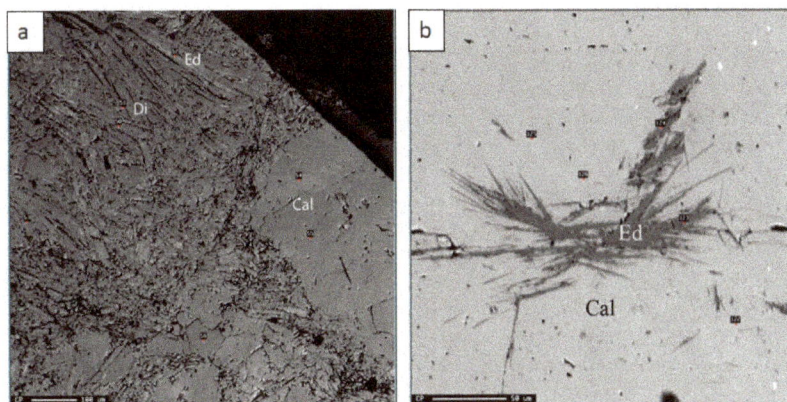

Figure 3. Secondary-electron image of serpentinite of the Pollino Massif showing (**a**) edenite, diopside, and calcite; (**b**) edenite crystals with both fibrous habit.

Results of the EMP analysis performed on selected fibrous crystals of edenite are shown in Table 4. Major element compositional range of this amphibole is as follows: SiO$_2$ = 51.264–54.293 wt%, CaO = 23.64–25.507 wt%, MgO = 16.332–17.680 wt%, Al$_2$O$_3$ = 0.259–2.709 wt%, and FeO$_{tot}$ = 1.257–2.852 wt% [57]. In addition, in the edenite crystals, low amounts of several trace elements, such as Mn, Cr, and Ni, are also present.

4.2. Asbestos Minerals in Serpentinites from Gimigliano-Mt. Reventino

The serpentinites show remnants of the original protogranular texture, which is inherited from their harzburgitic-lherzolitic protoliths [5,19]. The mineral assemblage is made of serpentine group minerals and magnetite ± tremolite-actinolite ± chlorite ± clinopyroxene ± Cr-spinel, and calcite [5,19] (Table 1). The serpentine group minerals, together with small magnetite grains, completely replaced the original olivine and orthopyroxene crystals that appear as pseudomorphic aggregates showing typical net-like and mesh textures [5,19]. According to Punturo et al. [19], clinopyroxene is in the rarely preserved holly-leaf shaped Cr-spinels that, in most cases, are quite completely retrogressed to magnetite and chlorite. Different vein systems, filled by serpentine group minerals, cross-cut the rock. In general, serpentine fibers may be oriented either perpendicular to the vein selvages ("cross" serpentine) or according to their elongation directions ("lamellar" serpentine). Minor calcite and talc flake aggregates or actinolite-tremolite fibers may occur within the serpentine matrix.

Table 3. Chemistry of selected fibrous tremolite and Mg-Fe-hornblende crystals in the serpentinites of the Pollino Massif.

N. Analysis	73	76	77	78	79	91	109	130	98	102
Oxides (wt%)	-	-	-	-	-	-	-	-	-	-
SiO_2	54.588	57.392	55.674	52.04	53.735	57.547	55.015	57.337	51.657	55.263
P_2O_5	0.031	n.d.	n.d.	0.01	0.016	0.028	0.024	0.057	n.d.	0.005
TiO_2	0.158	0.011	0.059	0.433	0.268	0.075	0.059	n.d.	0.482	0.065
Al_2O_3	2.798	0.479	1.559	5.282	3.543	1.369	2.509	n.d.	5.484	1.871
Cr_2O_3	0.224	0.007	0.092	0.425	0.486	0.009	0.225	n.d.	0.502	0.006
MnO	0.138	0.089	0.045	0.026	0.119	0.082	0.082	0.022	0.03	0.17
FeO	3.925	3.147	2.475	3.231	3.064	2.663	2.401	2.012	2.871	7.065
NiO	0.108	0.09	0.05	0.082	0.076	0.045	n.d.	n.d.	0.139	0.05
MgO	23.1	23.415	24.408	22.602	23.236	23.524	23.623	23.445	21.81	20.999
CaO	11.427	13.574	12.845	11.959	12.063	12.273	12.523	13.653	12.421	9.68
Na_2O	1.192	0.09	0.459	1.187	1.144	0.358	0.813	0.07	1.296	1.995
K_2O	0.002	n.d.	0.014	0.008	0.014	0.014	0.016	0.022	0.028	0.015
F	n.d.	0.093	n.d.	0.039	0.023	n.d.	0.037	0.032	n.d.	n.d.
Cl	n.d.	0.018	0.019	0.004	0.02	0.011	0.006	0.01	0.027	0.003
Sum	97.691	98.362	97.695	97.311	97.792	97.996	97.316	96.645	96.741	97.186
Final wt% values										
MnO	0.00	0.09	0.00	0.00	0.00	0.08	0.00	0.02	0.00	0.00
Mn_2O_3	0.15	0.00	0.05	0.03	0.13	0.00	0.09	0.00	0.03	0.19
FeO	0.00	0.09	0.00	0.00	0.00	0.54	0.00	0.35	0.00	0.00
Fe_2O_3	4.36	3.39	2.75	3.59	3.41	2.36	2.67	1.85	3.19	7.85
H_2O^+	2.14	2.14	2.17	2.05	2.10	2.18	2.15	2.18	2.05	2.15
Sum	100.28	100.88	100.15	99.74	100.26	100.42	99.76	99.03	99.12	100.14
Species	Tr	Tr	Tr	Tr	Tr	Tr	Tr	Tr	Mg-Fe-Hbl	Mg-Fe-Hbl
Formula Assignments T (ideally 8 apfu)										
Si	7.493	7.800	7.619	7.207	7.389	7.813	7.561	7.907	7.209	7.628
Al	0.453	0.077	0.251	0.792	0.574	0.185	0.406	0.000	0.791	0.304
Ti	0.016	0.001	0.006	0.000	0.028	0.000	0.006	0.000	0.000	0.007
Fe^{3+}	0.036	0.123	0.123	0.000	0.008	0.000	0.025	0.090	0.000	0.061
T subtotal	8.000	8.001	7.999	8.000	8.000	8.000	7.999	8.000	8.000	8.000
Formula Assignments C (ideally 5 apfu)										
Cr	0.024	0.001	0.010	0.047	0.053	0.001	0.024	0.000	0.055	0.001
Mn^{3+}	0.016	0.000	0.005	0.003	0.014	0.000	0.010	0.000	0.004	0.020
Fe^{3+}	0.414	0.225	0.160	0.374	0.344	0.241	0.251	0.102	0.335	0.755
Ni	0.012	0.010	0.006	0.009	0.008	0.005	0.000	0.000	0.016	0.006
Mg	4.534	4.744	4.819	4.452	4.581	4.711	4.715	4.820	4.429	4.219
C subtotal	5.000	5.001	5.000	5.000	5.000	5.000	5.000	4.965	5.001	5.001
Formula Assignments B (ideally 2 apfu)										
Mg	0.193	0.000	0.160	0.214	0.182	0.050	0.125	0.000	0.108	0.102
Ca	1.681	1.977	1.840	1.775	1.777	1.785	1.844	2.000	1.857	1.432
Na	0.126	0.023	0.000	0.011	0.041	0.094	0.031	0.000	0.035	0.467
B subtotal	2.000	2.000	2.000	2.000	2.000	1.999	2.000	2.000	2.000	2.001
Formula Assignments A (from 0 to 1 apfu)										
Ca	0.000	0.000	0.044	0.000	0.000	0.000	0.000	0.017	0.000	0.000
Na	0.191	0.000	0.122	0.308	0.264	0.000	0.186	0.019	0.316	0.067
K	0.000	0.000	0.002	0.001	0.002	0.002	0.003	0.004	0.005	0.003
A subtotal	0.191	0.000	0.168	0.309	0.266	0.002	0.189	0.040	0.321	0.070
Sum T,C,B,A	15.191	15.002	15.167	15.309	15.266	15.001	15.188	15.005	15.322	15.072

n.d. = not detected.

Table 4. Chemistry of selected fibrous edenite crystals in serpentinites samples from the Pollino Massif.

N. Analysis	50	51	57	58	61	63	68	69	70	77	78
Oxides (wt %)											
SiO_2	53.347	54.451	53.069	53.452	53.387	54.129	52.799	52.955	54.293	51.264	53.878
P_2O_5	0.009	0.020	0.022	0.057	0.013	0.014	0.006	0.015	0.002	0.010	0.051
TiO_2	0.006	0.006	0.017	0.025	0.000	0.012	0.041	0.012	0.021	0.002	0.023
Al_2O_3	0.958	0.259	1.529	0.858	0.747	0.486	2.709	0.785	0.382	1.869	0.388
Cr_2O_3	0.015	0.000	0.016	0.000	0.024	0.013	0.000	0.006	0.003	0.000	0.007
MnO	0.151	0.061	0.184	0.082	0.036	0.124	0.120	0.124	0.137	0.147	0.124
FeO	1.257	1.269	2.303	1.767	1.342	1.852	2.233	1.961	1.435	2.037	2.852
NiO	0.014	0.043	0.000	0.050	0.061	0.000	0.000	0.048	0.026	0.000	0.010
MgO	17.680	17.076	16.512	16.917	17.506	16.720	16.332	16.704	17.101	17.066	16.990
CaO	23.640	25.340	24.529	24.833	24.716	25.228	24.079	24.949	25.507	24.555	25.405
Na_2O	0.102	0.057	0.145	0.091	0.071	0.076	0.118	0.062	0.102	0.040	0.056
K_2O	0.046	0.028	0.036	0.037	0.007	0.000	0.014	0.026	0.000	0.018	0.005
F	0.000	0.000	0.014	0.000	0.000	0.024	0.020	0.000	0.000	0.013	0.000
Cl	0.023	0.016	0.073	0.028	0.014	0.011	0.016	0.002	0.013	0.013	0.007
Sum	97.24	98.63	98.43	98.19	97.92	98.68	98.48	97.65	99.02	97.02	99.80
Final wt % values											
Mn_2O_3	0.17	0.07	0.21	0.09	0.04	0.14	0.13	0.14	0.15	0.16	0.14
Fe_2O_3	1.40	1.41	2.56	1.96	1.49	2.06	2.48	2.18	1.60	2.26	3.17
H_2O^+	2.12	2.11	2.08	2.10	2.11	2.10	2.09	2.10	2.10	2.09	2.09
Sum	99.52	100.88	100.79	100.49	100.19	101.00	100.83	99.98	101.30	99.36	102.22
Species	Ed	Ed	Ed	Ed	Ed	Ed	Ed	Ed	Ed	Ed	Ed
Formula Assignments T (ideally 8 apfu)											
Si	7.573	7.647	7.485	7.552	7.553	7.607	7.421	7.532	7.605	7.351	7.516
P	0.001	0.001	0.001	0.003	0.001	0.001	n.d.	0.001	n.d.	0.001	0.003
Al	0.160	0.043	0.254	0.143	0.125	0.080	0.449	0.132	0.063	0.316	0.064
Ti	0.001	0.001	0.002	0.003	n.d.	0.001	0.004	0.001	0.002	n.d.	0.002
Fe^{3+}	0.149	0.149	0.258	0.209	0.159	0.218	0.126	0.233	0.168	0.244	0.333
T subtotal	7.884	7.841	8.000	7.910	7.838	7.907	8.000	7.899	7.838	7.912	7.918
Formula Assignments C (ideally 5 apfu)											
Cr	0.002	n.d.	0.002	n.d.	0.003	0.001	n.d.	0.001	n.d.	n.d.	0.001
Mn^{3+}	0.018	0.007	0.022	0.010	0.004	0.015	0.014	0.015	0.016	0.018	0.015
Fe^{3+}	n.d.	n.d.	0.014	n.d.	n.d.	n.d.	0.137	n.d.	n.d.	n.d.	n.d.
Ni	0.002	0.005	n.d.	0.006	0.007	n.d.	n.d.	0.005	0.003	n.d.	0.001
Mg	3.742	3.575	3.472	3.563	3.692	3.503	3.422	3.542	3.571	3.648	3.533
C subtotal	3.764	3.587	3.510	3.579	3.706	3.519	3.573	3.563	3.590	3.666	3.550
Formula Assignments B (ideally 2 apfu)											
Ca	2.000	2.000	2.000	2.000	2.000	2.000	2.000	2.000	2.000	2.000	2.000
B subtotal	2.000	2.000	2.000	2.000	2.000	2.000	2.000	2.000	2.000	2.000	2.000
Formula Assignments A (from 0 to 1 apfu)											
Ca	1.596	1.813	1.707	1.759	1.747	1.799	1.626	1.802	1.828	1.773	1.797
Na	0.028	0.016	0.040	0.025	0.019	0.021	0.032	0.017	0.028	0.011	0.015
K	0.008	0.005	0.006	0.007	0.001	n.d.	0.003	0.005	n.d.	0.003	0.001
A subtotal	1.632	1.834	1.753	1.791	1.767	1.820	1.661	1.824	1.856	1.787	1.813
Sum T,C,B,A	15.280	15.262	15.263	15.280	15.311	15.246	15.234	15.286	15.284	15.365	15.281

n.d. = not detected.

11

The X-ray diffraction analysis revealed that these rocks are mainly constituted of serpentine group minerals, followed by chlorite and tremolite [5,19] (Table 2). Calcite was detected less frequently and in low amounts. μ-Raman spectroscopy identified chrysotile, lizardite, and antigorite. In the spectral region associated with the structural bending characterization, serpentine group minerals are characterized by very similar μ-Raman spectra. In chrysotile, the characteristic ν5 (e) bending vibrations of the SiO_4 tetrahedra are shown at 388 and 344 cm^{-1}. Lizardite shows a very similar pattern to chrysotile [5], whereas antigorite displays a characteristic band occurring at 1042 cm^{-1} and an intense band at 683 cm^{-1}. The two bands observed at 378 and 634 cm^{-1} appear to be slightly shifted when compared with the same vibrations present in the chrysotile spectrum [5]. Similarly to the serpentinites of the Pollino Massif, the tremolite of Gimigliano-Mt. Reventino rocks is characterized by the presence only of Fe^{2+} detected by FT-IR analysis [57]. The Fe^{3+} presence is excluded because of the absence of the absorption bands at $\Delta = -50$ cm^{-1}. Further, the FT-IR and SEM-EDS analyses confirmed the presence of antigorite, chrysotile, and fibrous minerals, from the tremolite–actinolite series, in the samples from Mt. Reventino [20,21,58].

5. Discussion and Conclusions

Data presented in this study display a very similar mineralogical composition for both the considered serpentinites including serpentine group minerals, amphiboles, pyroxene, chlorite, talc, titanite, magnetite, and carbonates. However, compared to the serpentinites from Northern Calabria, the Pollino Massif serpentinites are characterized also by the presence of quartz, dolomite, and edenite.

Based on textural and mineralogical data, quartz and dolomite are always found in association with, talc likely suggesting their formation during a metasomatic event. According to Boschi et al. [59], in fact, serpentine minerals may easily alter to talc + dolomite assemblage as the following reaction:

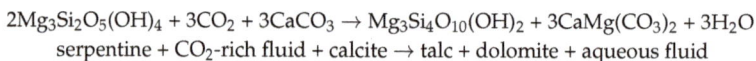

$$2Mg_3Si_2O_5(OH)_4 + 3CO_2 + 3CaCO_3 \rightarrow Mg_3Si_4O_{10}(OH)_2 + 3CaMg(CO_3)_2 + 3H_2O$$
$$\text{serpentine} + CO_2\text{-rich fluid} + \text{calcite} \rightarrow \text{talc} + \text{dolomite} + \text{aqueous fluid}$$

Instead, quartz may be thought as the result of the direct precipitation from migrating fluids that, as stated by Moore and Rymer [60], may become enriched with dissolved silica during the alteration of serpentine and other primary silicates of mafic and ultramafic rocks.

As regards edenite, the presence of such a mineral within the serpentinitic rocks from the ophiolitic sequence of Southern Apennines is documented here for the first time. Edenite is a double chain silicate mineral of the amphibole group with the following general formula: $NaCa_2(Mg,Fe)_5[Si_7AlO_{22}](OH)_2$ [61]. This is a rare mineral in the ophiolitic sequences although its presence has been documented in the oceanic serpentinites from the Mid-Atlantic Ridge and the Greater Antilles (Cuba, Dominican Republic) [62,63], where it testifies a medium to high metamorphism. As suggested by Bucher and Frey [64], the edenite formation is linked to the greenschist-amphibolite facies transition. In particular, such a mineral is produced by albite-consuming reactions in volcanic rocks, as albite + tremolite = edenite + 4 quartz, or in basic rocks, as olivine + labradorite + H_2O = ortopyroxene + edenite + spinel [61], and occurs during metamorphic events that promote systematic changes of the amphibole composition (from tremolite to edenite).

However, metasomatic processes not correlated to metamorphism can also be responsible for the formation of edenite or (more commonly) fluoro-edenite crystals. The F-edenite of Biancavilla, a village located in the etnean volcanic complex of Eastern Sicily (Southern Italy), is an example of an amphibole not metamorphic in origin. According to Comba et al. [65], the Sicilian F-edenite is found in voids and fractures of the benmoreitic lava covering the Mt. Calvario or in highly weathered pyroclastic products and scoriae that have been involved by metasomatizing hot fluids during volcanism or processes linked to it.

Regardless of process promoting edenite formation, it is worth noting that this amphibole, similarly to the more common asbestos minerals, may crystallize with fibrous habit and thus could represent a mineral harmful for human health, although edenite is currently not regulated by the Directive 2003/18/EC of the European Parliament either by the European Council of 27th March 2003. In addition, our data highlight that edenite crystals host different types of trace metals that could increase their toxicity as suggested by Bloise et al. [13] for the other asbestos minerals.

As a consequence, we believe that further detailed field and laboratory investigations are needed to improve our knowledge on edenite formation in the serpentinitic rocks of the Southern Apennines in order to better constrain the mode of occurrence of this "potentially harmful" mineral at the Pollino Massif.

Author Contributions: Conceptualization, G.R. and R.S.; methodology (EMPA measurements on edenite crystals), M.C.D. and M.P.; resources, G.R. and M.C.D.; data curation, M.C.D., G.R., and R.S.; writing—original draft preparation, M.C.D., M.P., G.R., and R.S.; supervision, G.R.

Funding: This research was funded by a Giovanna Rizzo grant, RIL2016.

Acknowledgments: The authors wish to thank the Centro Nacional de Microscopia Electrónica of the Universidad Complutense de Madrid (España) for the electron microprobe analyses. This work has received financial support from University of Basilicata.

Conflicts of Interest: The authors declare no conflict of interest.

References

1.   Belluso, E.; Compagnoni, R.; Ferraris, G. *Occurrence of Asbestiform Minerals in the Serpentinites of the Piemonte Zone, Western Alps*; Giornata di Studio in ricordo del Prof. Stefano Zucchetti; Politecnico di Torino: Torino, Italy, 1994; pp. 57–64.

2.   Groppo, C.; Rinaudo, C.; Cairo, S.; Gastaldi, D.; Compagnoni, R. Micro-Raman spectroscopy for a quick and reliable identification of serpentine minerals from ultramafics. *Eur. J. Mineral.* **2006**, *18*, 319–329. [CrossRef]

3.   Hendrickx, M. Naturally occurring asbestos in eastern Australia: A review of geological occurrence, disturbance and mesothelioma risk. *Environ. Geol.* **2009**, *57*, 909–926. [CrossRef]

4.   Vignaroli, G.; Ballirano, P.; Belardi, G.; Rossetti, F. Asbestos fibre identification vs. evaluation of asbestos hazard in ophiolitic rock mélanges, a case study from the Ligurian Alps (Italy). *Environ. Earth Sci.* **2013**, *72*, 3679–3698. [CrossRef]

5.   Bloise, A.; Punturo, R.; Catalano, M.; Miriello, D.; Cirrincione, R. Naturally occurring asbestos (NOA) in rock and soil and relation with human activities: The monitoring example of selected sites in Calabria (southern Italy). *Ital. J. Geosci.* **2016**, *135*, 268–279. [CrossRef]

6.   Virta, R.L. *Mineral Commodity Profiles: Asbestos*; USGS Circular 1255-KK; US Geological Survey (USGS): Reston, VA, USA, 2005; p. 56.

7.   Nichols, M.D.; Young, D.; Davis, G. *Guidelines for Geologic Investigations of Naturally Occurring Asbestos in California*; Special publication; California Geological Survey Public Information Offices: Los Angeles, CA, USA, 2002; p. 124.

8.   IARC Working Group on the Evaluation of Carcinogenic Risks to Humans, and World Health Organization. *Overall Evaluations of Carcinogenicity: An Updating of IARC Monographs*; World Health Organization: Geneva, Switzerland, 1987; Volume 1–42.

9.   Mossman, B.T.; Lippmann, M.; Hesterberg, T.W.; Kelsey, K.T.; Barchowsky, A.; Bonner, J.C. Pulmonary endpoints (lung carcinomas and asbestosis) following inhalation exposure to asbestos. *J. Toxicol. Environ. Health* **2011**, *14*, 76–121. [CrossRef] [PubMed]

10. Bloise, A.; Belluso, E.; Barrese, E.; Miriello, D.; Apollaro, C. Synthesis of Fe-doped chrysotile and characterization of the resulting chrysotile fibres. *Cryst. Res. Technol.* **2009**, *44*, 590–596. [CrossRef]

11. Bloise, A.; Barrese, E.; Apollaro, C.; Miriello, D. Flux growth and characterization of Ti- and Ni-doped forsterite single crystals. *Cryst. Res. Technol* **2009**, *44*, 463–468. [CrossRef]

12. Bloise, A.; Belluso, E.; Fornero, E.; Rinaudo, C.; Barrese, E.; Cappella, S. Influence of synthesis condition on growth of Ni-doped chrysotile. *Microporous Mesoporous Mater.* **2010**, *132*, 239–245. [CrossRef]

13. Bloise, A.; Barca, D.; Gualtieri, A.F.; Pollastri, S.; Belluso, E. Trace elements in hazardous mineral fibres. *Environ. Pollut.* **2016**, *216*, 314–323. [CrossRef] [PubMed]

14. Loreto, C.; Carnazza, M.L.; Cardile, V.; Libra, M.; Lombardo, L.; Malaponte, G.; Martinez, G.; Musumeci, G.; Papa, V.; Cocco, L. Mineral fiber-mediated activation of phosphoinositide-specific phospho-lipase c in human bronchoalveolar carcinoma-derived alveolar epithelial A549 cells. *Int. J. Oncol.* **2009**, *34*, 371–376.

15. Pugnaloni, A.; Giantomassi, F.; Lucarini, G.; Capella, S.; Bloise, A.; Di Primo, R.; Belluso, E. Cytotoxicity induced by exposure to natural and synthetic tremolite asbestos: An in vitro pilot study. *Acta Histochem.* **2013**, *115*, 100–112. [CrossRef] [PubMed]

16. Gueguen, E.; Doglioni, C.; Fernandez, M. On the post-25 Ma geodynamic evolution of the western Mediterranean. *Tectonophysics* **1998**, *298*, 259–269. [CrossRef]

17. Cello, G.; Mazzoli, S. Apennine tectonics in southern Italy: A review. *J. Geodyn.* **1999**, *27*, 191–211. [CrossRef]

18. Doglioni, C.; Gueguen, E.; Harabaglia, P.; Mongelli, F. On the origin of west-directed subduction zones and applications to the western Mediterranean. In *The Mediterranean Basins: Tertiary Extension within the Alpine Orogen*; Durand, B., Jolivet, L., Horváth, F., Séranne, M., Eds.; Special Publications Geological Society: London, UK, 1999; Volume 156, pp. 541–561.

19. Punturo, R.; Bloise, A.; Critelli, T.; Catalano, M.; Fazio, E.; Apollaro, C. Environmental implications related to natural asbestos occurrences in the ophiolites of the Gimigliano-Mount Reventino Unit (Calabria, Southern Italy). *Int. J. Environ. Res.* **2015**, *9*, 405–418.

20. Campopiano, A.; Olori, A.; Spadafora, A.; Rosaria Bruno, M.; Angelosanto, F.; Iannò, A.; Iavicoli, S. Asbestiform minerals in ophiolitic rocks of Calabria (southern Italy). *Int. J. Environ. Health Res.* **2018**, *28*, 134–146. [CrossRef] [PubMed]

21. Apollaro, C.; Fuoco, I.; Vespasiano, G.; De Rosa, R.; Cofone, F.; Miriello, D.; Bloise, A. Geochemical and mineralogical characterization of tremolite asbestos contained in the Gimigliano-Mount Reventino Unit (Calabria, south Italy). *J. Mediterr. Earth Sci.* **2018**, *1*, 5–15.

22. Knott, S.D. The Liguride Complex of southern Italy—A Cretaceous to Paleogene accretionary wedge. *Tectonophysics* **1987**, *142*, 217–226. [CrossRef]

23. Knott, S.D. Structure, kinematics and metamorphism of the Liguride complex, southern Apennines, Italy. *J. Struct. Geol.* **1994**, *16*, 1107–1120. [CrossRef]

24. Monaco, C.; Tansi, C.; Tortorici, L.; De Francesco, A.M.; Morten, L. Analisi geologico-strutturale dell'Unità del Frido al confine calabro-lucano (Appennino Meridionale). *Mem. Soc. Geol. Ital.* **1991**, *47*, 341–353.

25. Monaco, C.; Tortorici, L. Tectonic role of ophiolite-bearing terranes in building of the Southern Apennines orogenic belt. *Terra Nova* **1995**, *7*, 153–160. [CrossRef]

26. Cirrincione, R.; Monaco, C. Evoluzione tettonometamorfica dell'Unità del Frido (Appennino Meridionale). *Mem. Soc. Geol. Ital.* **1996**, *51*, 83–92.

27. Tortorici, L.; Catalano, S.; Monaco, C. Ophiolite-bearing melanges in southern Italy. *Geol. J.* **2009**, *44*, 153–166. [CrossRef]

28. Sansone, M.T.C.; Rizzo, G.; Mongelli, G. Petrochemical characterization of mafic rocks from Ligurian ophiolites, southern Apennines. *Int. Geol. Rev.* **2011**, *53*, 130–156. [CrossRef]

29. Laurita, S.; Prosser, G.; Rizzo, G.; Langone, A.; Tiepolo, M.; Laurita, A. Geochronological study of zircons from continental crust rocks in the Frido Unit (Southern Apennines). *Int. J. Earth Sci.* **2014**, *104*, 179–203. [CrossRef]

30. Dichicco, M.C.; Laurita, S.; Paternoster, M.; Rizzo, G.; Sinisi, R.; Mongelli, G. Serpentinite Carbonation for CO_2 Sequestration in the Southern Apennines: Preliminary Study. *Energy Procedia* **2015**, *76*, 477–486. [CrossRef]

31. Dichicco, M.C.; De Bonis, A.; Mongelli, G.; Rizzo, G.; Sinisi, R. Naturally occurring asbestos in the southern Apennines: Quick μ-Raman Spectroscopy identification as a tool of environmental control. In Proceedings of the 13th International Conference on Protection and Restoration of the Environment, Mykonos Island, Greece, 3–8 July 2016.

32. Dichicco, M.C.; De Bonis, A.; Mongelli, G.; Rizzo, G.; Sinisi, R. μ-Raman spectroscopy and X-ray diffraction of asbestos' minerals for geo-environmental monitoring: The case of the southern Apennines natural sources. *Appl. Clay Sci.* **2017**, *141*, 292–299. [CrossRef]

33. Dichicco, M.C.; Laurita, S.; Sinisi, R.; Battiloro, R.; Rizzo, G. Environmental and Health: The Importance of Tremolite Occurence in the Pollino Geopark (Southern Italy). *Geosciences* **2018**, *8*, 98. [CrossRef]

34. Dichicco, M.C.; Castiñeiras, P.; Galindo Francisco, C.; González Acebrón, L.; Grassa, F.; Laurita, S.; Paternoster, M.; Rizzo, G.; Sinisi, R.; Mongelli, G. Genesis of carbonate-rich veins in the serpentinites

at the Calabria-Lucania boundary (southern Apennines). *Rend. Online Soc. Geol. Ital.* **2018**, *44*, 143–149. [CrossRef]

35. Rizzo, G.; Laurita, S.; Altenberger, U. The Timpa delle Murge ophiolitic gabbros, southern Apennines: Insights from petrology and geochemistry and consequences to the geodynamic setting. *Period. Mineral.* **2018**, *87*, 5–20.

36. Mazzeo, F.C.; Zanetti, A.; Aulinas, M.; Petrosino, M.; Arienzo, I.; D'Antonio, M. Evidence for an intra-oceanic affinity of the serpentinized peridotites from the Mt. Pollino ophiolites (southern Ligurian Tethys): Insights into the peculiar tectonic evolution of the southern Apennines. *Lithos* **2017**, *284*, 367–380. [CrossRef]

37. Vitale, S.; Fedele, L.; Tramparulo, F.; Ciarcia, S.; Mazzoli, S.; Novellino, A. Structural and petrological analyses of the Frido Unit (southern Italy): New insights into the early tectonic evolution of the southern Apennines-Calabrian Arc system. *Lithos* **2013**, *168–169*, 219–235. [CrossRef]

38. Sansone, M.T.C.; Prosser, G.; Rizzo, G.; Tartarotti, P. Spinel peridotites of the Frido unit ophiolites (southern Apennines Italy): Evidence for oceanic evolution. *Period Miner.* **2012**, *81*, 35–59.

39. Sansone, M.T.C.; Tartarotti, P.; Prosser, G.; Rizzo, G. From ocean to subduction: The polyphase metamorphic evolution of the Frido unit metadolerite dykes (southern Apennine, Italy). Multiscale structural analysis devoted to the reconstruction of tectonic trajectories in active margins. *J. Virtual. Explor. Electron. Ed.* **2012**, *41*, 3.

40. Sansone, M.T.C.; Rizzo, G. Pumpellyite veins in the metadolerite of the Frido unit (southern Apennines-Italy). *Period. Mineral.* **2012**, *81*, 75–92.

41. Spadea, P. Calabria-Lucania ophiolites. *Boll. Geofis. Teorica Appl.* **1994**, *36*, 271–281.

42. Laurita, S.; Rizzo, G. Blueschist metamorphism of metabasite dykes in the serpentinites of the Frido Unit, Pollino Massif. *Rend. Online Soc. Geol. Ital.* **2018**, *45*, 129–135. [CrossRef]

43. Alvarez, W. Structure of the Monte Reventino greenschist folds: A contribution to untangling the tectonic-transport history of Calabria, a key element in Italian tectonics. *J. Struct. Geol.* **2005**, *27*, 1355–1378. [CrossRef]

44. Liberi, F.; Piluso, E. Tectonometamorphic evolution of the ophiolites sequences from Northen Calabrian Arc. *Ital. J. Geosci.* **2009**, *128*, 483–493.

45. Ogniben, L. Schema introduttivo alla geologia del confine calabro-lucano. *Mem. Soc. Geol. Ital.* **1969**, *8*, 453–763.

46. Ogniben, L. Schema Geologico della Calabria in base ai dati odierni. *Geol. Rom.* **1973**, *XII*, 243–585.

47. Piluso, E.; Cirrincione, R.; Morten, L. Ophiolites of the Calabrian Peloritani Arc and their relationships with the Crystalline Basement, Catena Costiera and Sila Piccola, Calabria, Southern Italy. *GLOM 2000 Excursion Guide-Book* **2000**, *25*, 117–140.

48. Pezzino, A.; Angi, G.; Fazio, E.; Fiannacca, P.; Lo Giudice, A.; Ortolano, G.; Punturo, R.; Cirrincione, R.; De Vuono, E. Alpine metamorphism in the aspromonte massif: Implications for a new framework for the southern sector of the Calabria-Peloritani Orogen, Italy. *Int. Geol. Rev.* **2008**, *50*, 423–441. [CrossRef]

49. Cirrincione, R.; Fazio, E.; Heilbronner, R.; Kern, H.; Mengel, K.; Ortolano, G.; Pezzino, A.; Punturo, R. Microstructure and elastic anisotropy of naturally deformed leucogneiss from a shear zone in Montalto (southern Calabria, Italy). *Geol. Soc. Lond. Spec. Publ.* **2010**, *332*, 49–68. [CrossRef]

50. Vitale, S.; Ciarcia, S.; Fedele, L.; Tramparulo, F. The Ligurian oceanic successions in southern Italy: The key to decrypting the first orogenic stages of the southern Apennines-Calabria chain system. *Tectonophysics* **2018**. [CrossRef]

51. Whitney, D.L.; Evans, B.W. Abbreviations for names of rock-forming minerals. *Am. Mineral.* **2010**, *95*, 185–187. [CrossRef]

52. Dichicco, M.C. Genesis of Carbonate-Rich Veins in the Serpentinites Outcropping at the Calabria-Lucania Boundary (Southern Apennines). Ph.D. Thesis, University of Basilicata, Potenza, Italy, 10 April 2018, unpublished.

53. Raudsepp, M.; Turnock, A.C.; Hawthorne, F.C.; Sherriff, B.L.; Hartman, J.S. Characterization of synthetic pargasitic amphiboles ($NaCa_2Mg_4M^{3+}Si_6Al_2O_{22}(OH, F)_2$; M^{3+} = Al, Cr, Ga, Fe, Sc, In) by infrared spectroscopy, Rietveld structure refinement, and ^{27}Al, ^{29}Si, and ^{19}F MAS NMR spectroscopy. *Am. Mineral.* **1987**, *72*, 580–593.

Fibers **2019**, *7*, 24

54. Ballirano, P.; Andreozzi, G.B.; Belardi, G. Crystal chemical and structural characterization of fibrous tremolite from Susa Valley, Italy, with comments on potential harmful effects on human health. *Am. Mineral.* **2008**, *93*, 1349–1355. [CrossRef]

55. Pacella, A.; Ballirano, P.; Cametti, G. Quantitative chemical analysis of erionite fibres using a micro-analytical SEM-EDX method. *Eur. J. Mineral.* **2016**, *28*, 257–264. [CrossRef]

56. Dichicco, M.C.; Laurita, S.; Mongelli, G.; Rizzo, G.; Sinisi, R. Environmental problems related to serpentinites in the Pollino Geopark (Southern Appennine). In *89° Congresso congiunto SGI-SIMP, Rendiconti Online*; Società Geologica Italiana: Catania, Italy, 2018; p. 578.

57. Skogby, H.; Rossman, G.R. The intensity of amphibole OH bands in the infrared absorption spectrum. *Phys. Chem. Miner.* **1991**, *18*, 64–68. [CrossRef]

58. Campopiano, A.; Olori, A.; Cannizzaro, A.; Ianno, A.; Capone, P.P. Quantification of tremolite in friable material coming from Calabrian ophiolitic deposits by infrared spectroscopy. *J. Spectroscop.* **2015**, *2015*, 974902. [CrossRef]

59. Boschi, C.; Früh-Green, G.L.; Escartín, J. Occurrence and significance of serpentinite-hosted, talc-and amphibole-rich fault rocks in modern oceanic settings and ophiolite complexes: An overview. *Ofioliti* **2006**, *31*, 129–140.

60. Moore, D.E.; Rymer, M.J. Talc-bearing serpentinite and the creeping section of the San Andreas fault. *Nature* **2007**, *448*, 795. [CrossRef]

61. Deer, W.A.; Howie, R.A.; Zussman, H.J. *Introduzione ai Minerali che Costituiscono le Rocce*; Zanichelli Editore S.P.A.: Modena, Italy, 1994.

62. Garcìa-Casco, A.; Torres-Roldán, R.L.; Milla, G.; Millán, P.; Schneider, J. Oscillatory zoning in eclogitic garnet and amphibole, Northern Serpentinite Melange, Cuba: A record of tectonic instability during subduction? *J. Metamorph. Geol.* **2002**, *20*, 581–598. [CrossRef]

63. Deschamps, F.; Guillot, S.; Godard, M.; Andreani, M.; Hattori, K. Serpentinites act as sponges for fluid-mobile elements in abyssal and subduction zone environments. *Terra Nova* **2011**, *23*, 171–178. [CrossRef]

64. Bucher, K.; Frey, M. *Petrogenesis of Metamorphic Rocks*; Springer-Verlag: Berlin, Germany, 2002; 341p.

65. Comba, P.; Gianfagna, A.; Paoletti, L. Pleural Mesothelioma Cases in Biancavilla are Related to a New Fluoro-Edenite Fibrous Amphibole. *Arch. Environ. Health Int. J.* **2003**, *58*, 229–232. [CrossRef]

fibers

Article

Obtaining an Artificial Aggregate from Cement-Asbestos Waste by the Melting Technique in an Arc-Resistance Furnace

Jerzy Witek [1], Bronisław Psiuk [1], Zdzisław Naziemiec [2] and Robert Kusiorowski [1,*]

[1] Institute of Ceramics and Building Materials, Refractory Materials Division in Gliwice, Toszecka 99, 44-100 Gliwice, Poland; j.witek@icimb.pl (J.W.); b.psiuk@icimb.pl (B.P.)

[2] Institute of Ceramics and Building Materials, Glass and Building Materials Division in Cracow, Cementowa 8, 31-983 Kraków, Poland; z.naziemiec@icimb.pl

* Correspondence: r.kusiorowski@icimb.pl or robert.kusiorowski@interia.pl; Tel.: +48-32-270-1939

Received: 20 December 2018; Accepted: 18 January 2019; Published: 24 January 2019

Abstract: Nowadays, asbestos waste still remains a serious problem. Due to the carcinogenic properties of asbestos, which are related to its fibrous structure, the exposure to asbestos mineral and asbestos-containing materials (ACM) causes dangerous health effects. This problem can be solved by recycling techniques, which allow the re-use of neutralized asbestos waste, instead of disposing it in special landfills. The article presents the results of research aimed at investigating the possibility of obtaining aggregates from asbestos waste by the fusion process in the electric arc-resistance process. A mixture of ACM with selected fluxes was were melted and then cast to form a grain of aggregates. The chemical composition of the material was determined before and after the melting process. Scanning electron microscopy (SEM) and X-ray diffraction (XRD) were applied to evaluate the effects of the fusion process. The main properties of the obtained aggregate were also measured. The results confirmed that the fibrous structure of asbestos was destroyed in the obtained material, which can be successfully used for the production of artificial aggregates.

Keywords: asbestos-containing materials; recycling; thermal treatment; melting process; aggregates

1. Introduction

Asbestos is the commercial name of a specific group of minerals, which occur in nature and belong to hydrated silicates containing mainly magnesium, sodium, calcium, and iron. These minerals are characterized by a specific crystalline structure, which in the microscopic examination is visible as a fibrous form. There are six asbestos minerals: Chrysotile, crocidolite (a fibrous form of riebeckite), amosite (a fibrous form of grunerite), anthophyllite, tremolite, and actinolite.

The characteristic fibrous structure of asbestos allows easy longitudinal splitting of thicker fibres, whereas transverse splitting of fibres is quite difficult, and results in their exceptional elasticity and high mechanical strength. The specific physical and chemical properties of asbestos, including its resistance to high temperatures and caustic substances, as well as the low costs of its obtaining (mining) has resulted in it commonly used in many branches of economy in the past. Due to its non-biodegradability and resistance to the attack of various agents, asbestos is a material difficult to destroy and, once placed in the environment, it can remain there for hundreds of years.

Only one type of asbestos belongs to the serpentine group, i.e. chrysotile (white asbestos), which is extracted and used in the largest quantities. The remaining minerals belong to a group of amphiboles, among which only two types are important for industrial practice: Amosite asbestos (brown asbestos) and crocidolite asbestos (blue asbestos) [1–4].

Asbestos was one of the most popular and cheapest raw materials used for the production of building materials. It was also applied in the machinery and shipbuilding industry. At the peak of its popularity, there were more than 3000 different asbestos-containing products on the market [5]. The major ones included: (a) Cement-asbestos products made from chrysotile and amphibole asbestos (roof slates, pressure pipes, stone cladding, and façade panels)—the largest group of asbestos products; (b) insulating products used to insulate steam boilers, heat exchangers, tanks, pipes, as well as clothing and fire-proof fabrics; (c) sealing products, which include cardboard, woven sealants, asbestos-rubber slabs; (d) abrasive products, such as: Friction linings and brake bands; (e) hydro-insulating products: Asphalt binders, sealing putties, refined road asphalts [3,6].

In many countries, the production of asbestos-containing materials has been limited and considerably reduced. As asbestos proved to have carcinogenic properties, for many years it has been included in Group 1 by the International Agency for Research and Cancer (IARC) [7]. The pathogenic effect of asbestos results from inhaling asbestos fibres suspended in the air. Inhalation of asbestos fibres does not cause immediate pathological changes—they are visible 10–60 years after the first contact. Exposure to asbestos dust can be a cause of many diseases, including among others: Asbestos pneumoconiosis (asbestosis), pleural changes, lung cancer, and mesotheliomas [7–9].

After discovering the harmful effects of asbestos fibres on human health, the extraction of asbestos minerals and manufacturing of asbestos-based products was gradually limited. Finally, many countries introduced regulations that not only banned the manufacturing of asbestos-containing products, but also made it mandatory to remove such products from the economy, in particular, from building facilities, and store them at controlled landfill sites. For example, in Poland, the Council of Ministers adopted a "programme for removal of asbestos and asbestos-containing products on the territory of Poland" [10], according to which the entire process of removal should have finished by 2032.

This is only a partial solution as it does not eliminate asbestos and its fibrous structure completely, but merely isolates it from society. This poses a secondary risk for the surrounding environment. Therefore, it seems necessary to search for methods based on complete degradation of asbestos and material recycling of the product obtained after treatment. The resolution of the European Parliament of 14 March 2013 [11] makes it clear that "delivering asbestos waste to landfills would not appear to be the safest way of definitively eliminating the release of asbestos fibres into the environment, and therefore, it would be far preferable to opt for asbestos inertisation plants." It further points out that, "as regards the management of asbestos waste, measures must also be taken to promote and support research into, and technologies using, eco-compatible alternatives, and to secure procedures, such as the inertisation of waste-containing asbestos, to deactivate active asbestos fibres and convert them into materials that do not pose public health risks".

The asbestos problem is widely known and raises concerns all over the world [12,13]. In Poland alone, the scale of this problem is particularly large. According to estimates, approximately 15.5 million tonnes of asbestos-containing materials are accumulated in Poland, of which 14.9 million tonnes (i.e. 96%) are cement-asbestos products [10]. The progress in the disposal of asbestos waste by landfilling in Poland seems to be insufficient. According to latest data [14], only slightly more than 6.2 million tonnes of asbestos-containing waste was inventoried in Poland, of which approximately 0.9 million tonnes was disposed of (through storage).

The process of neutralization, utilization, and recycling of asbestos has been the subject of numerous investigations. Most of the methods used involve destroying the harmful structure of asbestos fibres and turning the material into one that can be potentially reused. The vast majority of the described methods concern the thermic method of destroying the fibrous structure of asbestos [15–37]. There is a multitude of solutions proposed in this respect, depending on the source of heat and the manner of heating (conventional, microwave, plasma etc.), or the way asbestos fibres are destroyed (controlled recrystallization, vitrification, melting etc.). There are also reports of chemical [38–43] or mechanochemical methods [44,45]. Detailed information on various methods of dealing with asbestos waste is provided by review studies, which have appeared in recent years [46–48]. One of the proposed solutions is also a method of asbestos waste disposal by a melting process in a laboratory arc-resistance electric furnace. This issue was studied by the authors in their previous work [49]. The obtained results showed that the fibrous structure of asbestos, contained in cement-asbestos waste was completely changed and destroyed. As a result of this treatment, new mineral phases without dangerous properties were formed. The obtained results indicate that the melting process is an interesting method of neutralizing hazardous asbestos materials. On the other hand, finding new ways of managing the previously processed waste will allow for effective waste management. One such potential directions seems to be the production of artificial aggregates from asbestos waste. The problem of artificial aggregates made from various waste materials is common. Artificial aggregates from industrial by-products (fly ash, mining residues, sewage sludge etc.) are increasingly used on a large scale. They offer a solution to two significant problems of sustainable development, i.e. they may protect the environment against pollution and prevent natural resources from being depleted [50,51]. This work extends the scope of the authors' previous research.

The aim of this work was to study the melting of asbestos-containing waste in an electric arc-resistance furnace, with the participation of specially-selected mineral additives, which enable the obtaining of a useful final material—artificial construction aggregate. In addition to laboratory tests that check the destruction of asbestos forms in the melting process, the usability of the material obtained was also assessed in terms of the requirements to be fulfilled by construction aggregates.

2. Materials and Methods

The following materials were selected for the study: (1) Cement-asbestos waste board (in Poland, it is commonly known as "eternit") was used as a representative of asbestos-containing material (ACM)—it was very popular as a construction material for various buildings and systems in the second half of the twentieth century. It now represents the biggest group of ACM accumulated on the territory of Poland. It is assumed that cement-asbestos waste accounts for more than 90% of asbestos products accumulated in Poland; (2) a mixture of soda (sodium carbonate) and potassium feldspar were used as fluxes applied in the study; both raw materials were commercially available and had technical grades of purity. Based on previous investigations [49], these raw materials were selected for cement-asbestos melting, due to their relatively high efficiency, widespread occurrence, and the absence of hazardous substances formed during high-temperature treatment.

A mixture of ACMs, with the above-mentioned fluxes (85 wt% of ACM, 10 wt% of soda, and 5 wt% of feldspar) was melted for 100 minutes in an electric arc-resistance furnace and, then, cast into a ceramic mould to form a material for aggregates. Trials were carried out in a laboratory arc-resistance furnace (100 kVA autotransformer, 35 dm^3 working chamber) coupled with two graphite electrodes (5-cm diameter, mounted in a "V" system). The melting process was carried out in a furnace with a magnesia lining.

The chemical analysis of the raw materials and melted product was performed by X-ray fluorescence (XRF; PANalytical, Almelo, The Netherlands), using a Panalytical Magix PW-2424 spectrometer, following the procedures contained in the PN-EN ISO 12677:2011 standard [52]. To obtain homogeneous samples, the fused cast-bead method was applied. The chemical analysis data was supplemented by the content of volatile components, measured by calcination at 1025 °C and expressed as a value of loss on ignition (LOI). A thermo-gravimetric analysis (TG-DTG; Netzsch, Selb, Germany), combined with an evolved gas analysis (EGA) was performed in an alumina crucible, using an STA 409PC NETZSCH thermal analyser with quadrupole mass spectrometer QMS 403C Aëolos in the 25–1450 °C temperature range, with a heating rate of 5 °C·min^{-1} and a sample of 130 mg. The tests were carried out in a synthetic air atmosphere. The phase composition of the raw cement-asbestos waste, and the obtained melted product, was determined by powder X-ray diffraction (XRD; PANalytical, Almelo, The Netherlands). Analyses were conducted using a PANalytical X'pert Pro diffractometer (CuKα radiation, Ni filter, 40 kV, 30 mA, X'Celerator detector). A mineralogical quantitative phase analysis for the melted product was performed using the Rietveld method. The microstructure of raw samples, and the destruction of the fibrous nature of asbestos in ACM after melting, was observed by a scanning electron microscope (Mira III; Tescan, Brno, Czech Republic) in combination with the Energy Dispersive Spectroscopy (EDS) system, with AZtec Automated software (Oxford Instruments, Abingdon, UK).

The main properties of the obtained melted product, being potentially valuable material for aggregates, were checked by the procedures described below. The molten material was crushed in a single-toggle jaw crusher (Makrum, L44.41). Quality tests for the obtained artificial aggregate were carried out in accordance with PN-EN 12620 [53] and PN-EN 13043 [54] standards. The following parameters were determined for the obtained artificial aggregate: (a) Particle size distribution (according to PN-EN 933-1 [55]), (b) flakiness index (according to PN-EN 933-3 [56]), (c) crushed aggregate particle content (according to PN-EN 933-5 [57]), (d) grain density and absorbability (according to PN-EN 1097-6 [58]), (e) resistance to fragmentation by the Los Angeles test (according to PN-EN 1097-2 [59]), and (f) freezing resistance of the aggregate (according to PN-EN 1367-1 [60]).

3. Results and Discussion

3.1. Chemical and Phase Composition of the ACM

The chemical and phase compositions of raw cement-asbestos waste have been presented in Table 1 and Figure 1. Table 1 also contains chemical compositions of raw materials used for correcting the composition of asbestos-cement in the melting process.

Table 1. Results of raw materials' chemical analyses (values in wt%).

Sample	SiO_2	TiO_2	Al_2O_3	Fe_2O_3	MgO	CaO	Na_2O	K_2O	LOI
cement-asbestos (raw)	19.3	0.2	3.9	2.9	5.8	41.8	<0.1	0.4	25.1
potassium feldspar	66.6	0.1	17.7	0.1	0.1	0.5	2.7	11.7	0.5
soda (sodium carbonate)	-	-	-	-	-	-	58.0	-	41.5

LOI = loss on ignition.

In the case of the ACM sample, both chemical and phase compositions are typical of asbestos-cement products, in which the major components, apart from asbestos, included sand, Portland cement, and gypsum. The chemical composition is, therefore, dominated by CaO (ca 42 wt%), SiO_2 (ca 19 wt%) and MgO (ca 6 wt%). The chemical composition and the manner of asbestos binding in the cement matrix are also related to high values of loss on ignition (ca 25 wt%), which result from the thermal decomposition of particular mineral components (point 3.2).

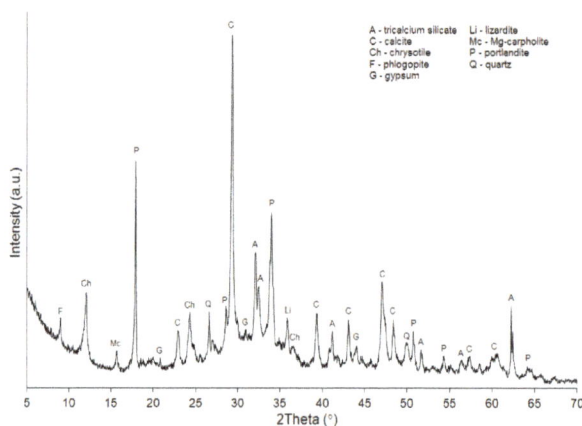

Figure 1. X-ray diffraction (XRD) pattern of a cement-asbestos waste board.

The phase composition (Figure 1) of the tested ACM was typical of this group of asbestos materials ad comparable with the results presented elsewhere [61]. One type of asbestos minerals in the ACM sample, i.e. chrysotile (ICDD-PDF 00-027-1275), was identified. This is confirmed by the main asbestos X-ray reflections at around 11–12° and 24–25° 2-Theta. Due to the presence of cementitious matrix, calcium carbonate (calcite; ICDD-PDF 01-072-1937) and calcium hydroxide (portlandite; ICDD-PDF 00-004-0733) were detected in the tested material. Calcite is present, due to cement matrix long-term weathering and gradual carbonation by atmospheric CO_2. This is a slow process, during which calcium hydroxide and/or hydrated phases of cement hydration (CSH phase) react with carbon dioxide. The CSH phase usually displays a low degree of crystallinity, so it is hardly visible on XRD patterns. The presence of X-ray amorphous compounds in the ACM sample is only visible after increasing the halo at 25–35° 2-Theta. Tricalcium silicate (ICDD-PDF 04-011-1393) was also identified as an unreacted phase of cement clinker. Trace amounts of lizardite (ICDD-PDF 01-087-2052), phlogopite (ICDD-PDF 00-010-0495), and magnesiocarpholite (ICDD-PDF 00-027-0303) were identified in the aged asbestos-cement material.

3.2. ACM Thermal Analysis Characterization

The results of the asbestos material thermal analysis (Figure 2) indicate that an approximately 10 wt% mass loss, at temperatures up to ca 300 °C, is related to the loss of physically bound water and dehydration of the CSH phase of bound cement and gypsum, which was put into the system as a binding time regulator. At a temperature of ca 450 °C, one can observe a visible effect of portlandite (calcium hydroxide) thermal decomposition, which is accompanied by a 2.5 wt% mass loss, related to water release according to the reaction: $Ca(OH)_2 \rightarrow CaO + H_2O$. On the other hand, within a temperature range of 500–800 °C, a considerable mass loss (ca 14 wt%), combined with a release of both water and carbon dioxide can be observed, which is confirmed by peaks recorded on the mass spectra of gases released from the sample in this temperature range. The wide temperature range indicates the overlapping of several processes during the thermal decomposition of the ACM sample. One of them, of course, is the decomposition of calcium carbonate (calcite) according to the following reaction: $CaCO_3 \rightarrow CaO + H_2O$. Stepkowska et al. [62] reported that, in this temperature range, the rest of the absorbed water escapes from aged cement pastes.

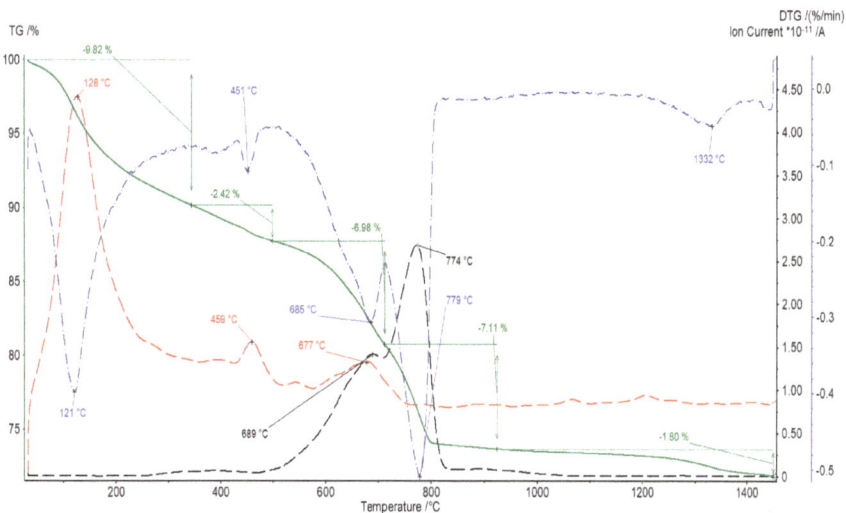

Figure 2. Thermo-gravimetric analysis (TG-DTG) results combined with evolved gas analysis (EGA) analysis (H_2O—red line; CO_2—black line) for a cement-asbestos waste board.

Another phenomenon in this range of temperatures, which seems important from the point of view of this work, is also the dehydroxylation of minerals, like chrysotile (white asbestos) and lizardite, combined with their thermal decomposition and the formation of new mineral phases. The thermal decomposition of these minerals with the formation of new mineral phases, like forsterite (Mg_2SiO_4) and/or enstatite ($MgSiO_3$), was thoroughly studied in the past few years [30,32,63]. The above transformations could be simplified and represented by the following overall reaction: $Mg_3(OH)_4Si_2O_5 \rightarrow Mg_2SiO_4 + MgSiO_3 + 2H_2O$, which allows for the thermal transformation of asbestos fibers into new minerals, characterized by complete recrystallization. A slight loss of mass at ca 1300 °C (1.8 wt%) is most probably related to the decomposition of sulphates, mainly $CaSO_4$, which are present in the tested ACM material in the form of gypsum.

3.3. Characterization of the Melted Product

The process of melting ran smoothly and was accompanied by a release of certain amounts of gases. This effect was related to the presence of compounds in the asbestos-cement material, which thermally decomposed at high temperatures, with a release of gaseous products (H_2O, CO_2, SO_2). The yield of melt in relation to the mass of feed subjected to melting, reached approximately 70%. However, if the loss of ignition of the ACM material is taken into account, it can be assumed that this yield was practically 100% and the feed was completely melted. During the experiment the measured power consumption (the electricity consumed by the furnace), in relation to the mass of the obtained melt was approximately 2.2 kWh·kg^{-1}. However it should be taken into consideration that the mass of the lining of the furnace is significantly higher (nearly 10 times) compared to the melted raw material charge. A significant amount of energy is consumed to warm up the furnace ceramic lining. When the process is scaled up, the thermal and electric consumption will be lower. It can be assumed that the energy consumption of the proposed method has similar or even lower values than in other ACM thermal treatment processes (ranging from 0.5 to 1.5 kWh·kg^{-1} [48]), so it can be considered to be economically competitive.

Table 2. Results of the chemical analysis and quantitative X-ray phase analyses of the obtained aggregate (values in wt%).

Sample	SiO₂	TiO₂	Al₂O₃	Fe₂O₃	MgO	CaO	Na₂O	K₂O	LOI
melted product (aggregate)	27.1	0.2	6.0	3.6	7.3	51.1	0.3	3.1	0.2
	calico-olivine		larnite		periclase		brownmillerite		Na-aluminate
	19.8		68.3		4.2		1.1		6.6

LOI = loss on ignition.

An analysis of the chemical composition (Table 2) and phase composition (Figure 3) indicates that the main phase in the product of melting of ACM with an addition of sodium carbonate and feldspar, is dicalcium silicate in the form of larnite crystals (β-Ca_2SiO_4; ICDD-PDF 01-077-0388) and calcio-olivine (γ-Ca_2SiO_4; ICDD-PDF 04-010-9508).

The quantitative phase composition analysis demonstrated that their contents were 68.3 and 19.8 wt%, respectively. Moreover, approximately 7% of sodium aluminate ($NaAlO_2$, ICDD-PDF 04-006-9358), and a small amount of brownmillerite (Ca_2FeAlO_5, ICDD-PDF 04-011-5939), were identified. An important observation resulting from the analysis of the phase composition revealed that the melted sample did not contain any asbestos, as the characteristic reflexes from chrysotile were not found on the diffraction pattern. On the other hand, approximately a 4 wt% content of magnesium oxide, in the form of periclase (MgO, ICDD-PDF 04-012-6481), was observed. The main component in the raw ACM sample, that was responsible for introducing magnesium compounds into the system, was asbestos fibre—chrysotile. As demonstrated by the research of Belardi and Piga [28], chrysotile asbestos in the presence of calcium compounds, for example from the thermal decomposition of portlandite or calcite, is subject to thermal decomposition to dicalcium silicate, with a release of free magnesium oxide according to the reaction: $Mg_3(OH)_4Si_2O_5 + 4CaO \rightarrow 2Ca_2SiO_4 + 3MgO + 2H_2O$. Therefore, it can be concluded that the presence of MgO in the melting product, indirectly confirms the decomposition of asbestos in the material.

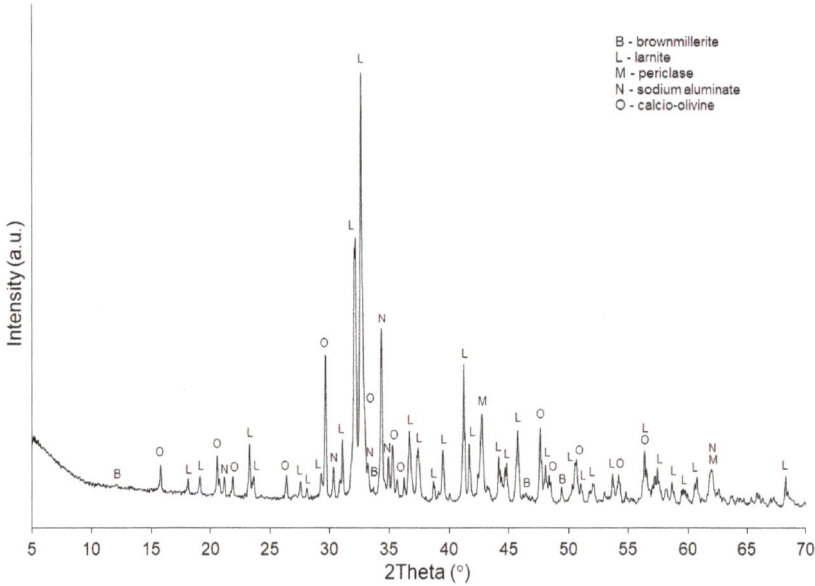

Figure 3. XRD pattern of the material obtained after the melting process.

3.4. SEM Characterization of Materials before and after Treatment

The most important feature, from the point of view of the aim of this work, was the presence of asbestos fibres in the cement-asbestos sample. The observations were conducted both after, and before, melting for comparison purposes. An Scanning electron microscopy (SEM)/EDS examination of the micro-area revealed the presence of chrysotile asbestos in the ACM material (Figure 4). On the other hand, the solidified material, obtained after melting, did not contain any fibrous forms (Figure 5). Microscopic observations of the ACM sample, subjected to melting, revealed that this material is characterized by a non-homogenous and fine-crystalline structure. The EDS measurement showed that the sample is mainly built of calcium silicate (represented by X-ray-identified larnite (β-C$_2$S) and calcio-olivine (γ-C$_2$S)). In the SEM photograph (Figure 5a), there are visible voids after asbestos utilization. These voids have the same chemical composition as the environment, which confirms asbestos utilization. Chrysotile destruction is also confirmed by the fact that magnesium, which is a chrysotile component in the sample before melting, occurs in the form of isolated MgO islands in the sample after melting. The identified contents of periclase were clearly separated from the remaining part of the material. Therefore, the conducted SEM/EDS analysis confirmed the complete destruction of asbestos fibres during the melting process.

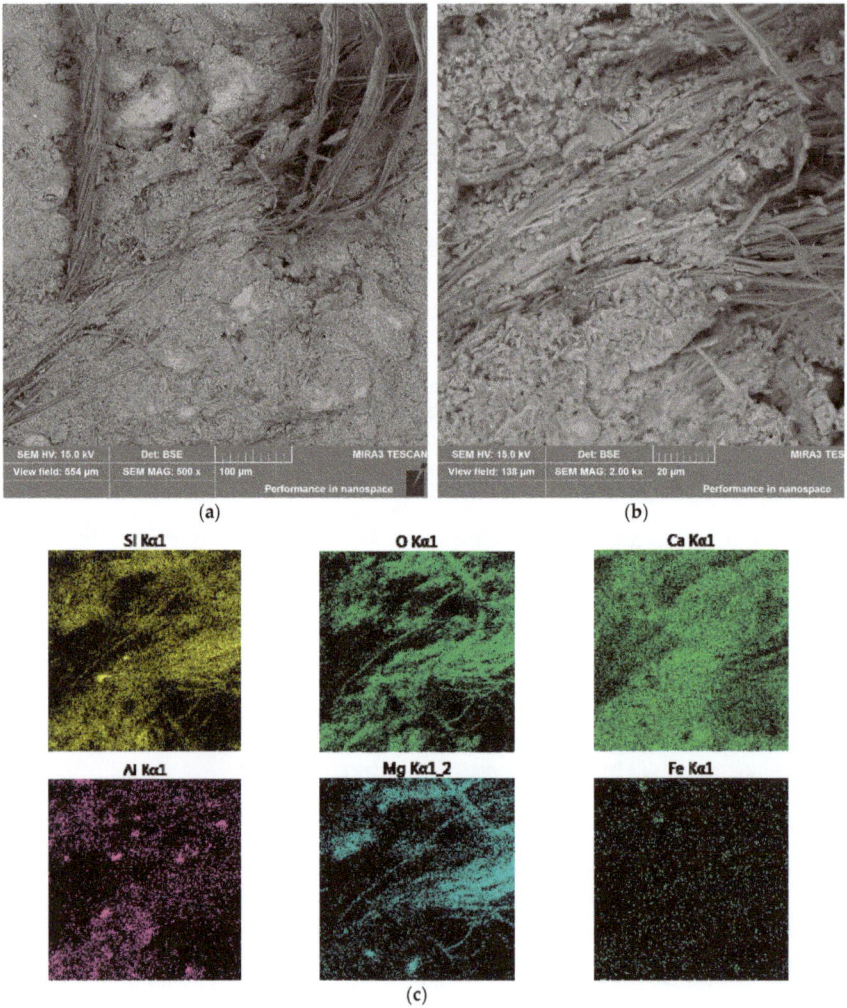

(a) **(b)**

(c)

Figure 4. Scanning electron microscopy (SEM) images of the asbestos-containing materials (ACM) sample: Surface; 500× (**a**); fracture; 2000× (**b**) and corresponding fracture Energy Dispersive Spectroscopy (EDS) elemental maps (**c**).

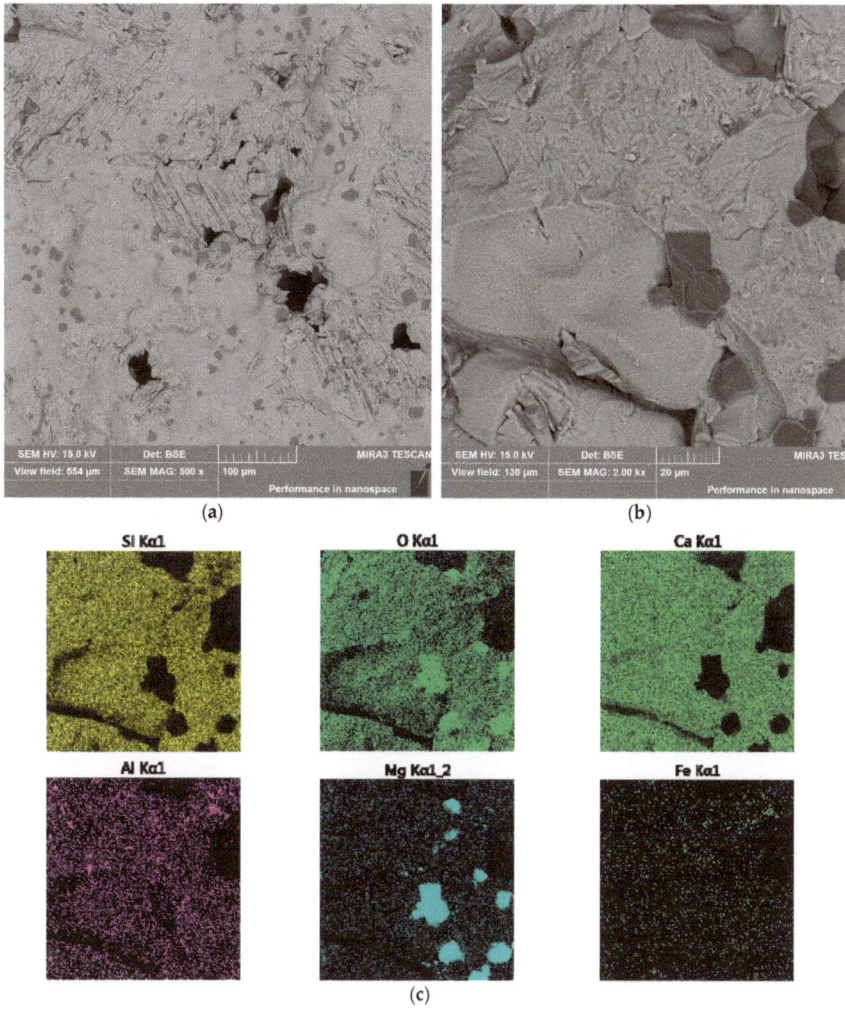

Figure 5. SEM images of the ACM sample after thermal treatment: surface; 500× (**a**); fracture; 2000× (**b**) and corresponding fracture EDS elemental maps (**c**).

3.5. Properties of the Obtained Artificial Aggregate

For the artificial aggregate obtained in the process of ACM melting with selected fluxes (composition: 85 wt% of ACM, 10 wt% of soda and 5 wt% of feldspar), the major quality characteristics were determined according to the the following standards: PN-EN 12620 and PN-EN 13043. The results of investigations into the basic functional properties have been presented in Table 3 and Figure 6.

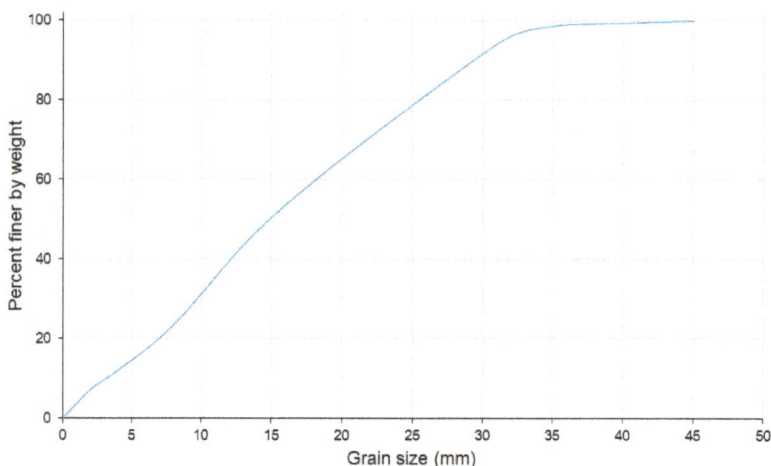

Figure 6. Particle size distribution of the artificial aggregate obtained after crushing the material in a jaw crusher.

Table 3. Main properties of the artificial aggregate obtained from asbestos-containing material (ACM).

Property	Abbreviation	Result	Standard
flakiness index	FI	$18 \pm 2\%$	PN-EN 933-3
particle density	ρ_a	$3.20 \pm 0.06 \text{ mg·m}^{-3}$	PN-EN 1097-6
	ρ_{rd}	$3.18 \pm 0.06 \text{ mg·m}^{-3}$	
	ρ_{ssd}	$3.20 \pm 0.06 \text{ mg·m}^{-3}$	
water absorption	WA_{24}	$0.10 \pm 0.05\%$	PN-EN 1097-6
percentage of crushed and broken surfaces	fraction 4/16 mm	$C_{100/0}$	PN-EN 933-5
resistance to freezing and thawing	fraction 4/8 mm	$3.2 \pm 1.0\%$	PN-EN 1367-1
	fraction 8/16 mm	$1.7 \pm 0.6\%$	
resistance to fragmentation (Los Angeles)	LA	$12 \pm 1\%$	PN-EN 1097-2

The investigated artificial aggregate obtained in the process of asbestos-cement melting is characterized—with a few exceptions—by good physical and mechanical properties. The grains of the aggregate, subjected to mechanical crushing, pass through a sieve with 45-mm mesh, more than 60 wt%, which have a size below 20 mm. Of course, the grain size distribution of the obtained artificial aggregate—depending on the needs—can be shaped by subjecting it to crushing and sieving in appropriately selected equipment. The artificial aggregate obtained in the process of ACM melting is characterized by a favourable size of grains. The flakiness FI of 18% results directly from the manner of aggregate preparation, as jaw crushers allow obtaining aggregates, with a relatively high content of shapeless grains. This result (18% of shapeless grains) allows the obtained aggregate to be classified into higher grades. According to the standards, it is assumed that the content of flat grains, required from the highest class aggregates, should be < 10% (according to PN-EN 13043) or < 15% (according to PN-EN 12620). On the other hand, the content of shapeless grains in the lowest grades of aggregates, can reach up to 50%. The obtained aggregate is characterized by a relatively high density of the grains. Grain density for natural aggregates usually ranges from 2.6 to 2.7 mg·m^{-3}, and in the case of the artificial aggregate obtained from ACM, reaches ca 3.2 mg·m^{-3}. It is worth noting that the obtained values of apparent, dry, and saturated-surface-dry (SSD) densities are very similar. For comparison—the highest values of bulk density among natural aggregates are obtained for basalts, reaching ca 3.0 mg·m^{-3}. Slight differences in density values result from the very low value of water absorption, which was approximately 0.1%. Such a low value is the effect of melting and subsequent solidification of the waste into a non-porous block. As the artificial aggregate was obtained by breaking and crushing larger, homogenous blocks, the $C_{100/0}$ category should be assumed, which corresponds to 100% content of totally, partially crushed, or broken grains with a complete absence of round ones. Resistance to crushing and fragmentation sd determined in the *Los Angeles testing* drum is very high, similar to that of aggregates obtained from magma rocks. For the highest categories, the LA index, according to PN-EN 13043 and PN-EN 12620, should be lower than 15%. For the aggregate obtained from ACM, the achieved value of LA index reached 12%. The aggregate's freezing resistance was relatively low and resulted from the material's low absorbability. In the case of the best natural aggregates obtained from magma and sedimentary rocks, the loss of mass of the aggregate, subjected to freezing and thawing, does not usually exceed 1–2%. In the case of the aggregate based on melted ACM, the value of resistance to freezing reached approximately 2–3%.

4. Conclusions

The obtained results demonstrated that the fibrous nature of asbestos was completely destroyed by melting the material with selected fluxes. The process of melting with appropriate additives allowed a new material to be obtained, which can be successfully used for the production of artificial aggregates. The investigated aggregates fulfil the requirements for different levels of categories, as defined in PN-EN 12620 Aggregates for concrete and PN-EN 13043 Aggregates for bituminous mixtures and surface treatments for roads, airfields, and other trafficked areas. In the future they can be used in road construction as well as in the production of concrete, when the processing of asbestos waste is allowed as the only possible method of its neutralization.

Author Contributions: J.W. conceived the research, performed technological measurements, and analysed the results; Z.N. determined the properties of the aggregate, B.P. performed SEM/EDS analysis, R.K. analysed the results and wrote the manuscript.

Funding: This work is the result of statutory activity at the Institute of Ceramics and Building Materials—Refractory Materials Division supported by the Polish Ministry of Science and Higher Education.

Conflicts of Interest: The authors declare no conflict of interest.

References

1. Virta, R.L. *Mineral Commodity Profiles—Asbestos*; Circular 1255-KK; U.S. Geological Survey: Reston, VA, USA, 2005.
2. Gualtieri, A.F. (Ed.) *Mineral Fibers: Crystal Chemistry, Chemical-Physical Properties, Biological Interaction and Toxicity*; European Mineralogical Union: London, UK, 2017; Volume 18, ISBN 978-0903056-65-6.
3. Obmiński, A. *Asbestos in Buildings*; Building Research Institute: Warsaw, Poland, 2017; ISBN 978-83-249-8477-0.
4. Sporn, T.A. Mineralogy of asbestos. In *Malignant Mesothelioma*; Tannapfel, A., Ed.; Springer: Berlin, Germany, 2011; pp. 1–11. ISBN 978-3-642-10861-7.
5. Harris, L.V.; Kahwa, I.A. Asbestos: Old foe in 21st century developing countries. *Sci. Total Environ.* **2003**, *307*, 1–9. [CrossRef]
6. Pyssa, J.; Rokita, G.M. The asbestos—Occurence, using and the way of dealing with asbestic waste material. *Min. Res. Manag.* **2007**, *23*, 49–61.
7. IARC. *Overall Evaluations of Carcinogenicity: An Updating of IARC Monographs Volumes 1 to 42*; IARC: Lyon, France, 1987; 440p.
8. IARC. *IARC Monographs on the Evaluation of Carcinogenic Risks to Humans. Volumes 81 Man-Made Vitreous Fibres*; IARC: Lyon, France, 2002; 433p.
9. IARC. *IARC Monographs on the Evaluation of Carcinogenic Risks to Humans. Volumes 100C Arsenic, Metals, Fibres and Dust*; IARC: Lyon, France, 2012; 527p.
10. *Annex to the Resolution No. 39/2010 of the Council of Ministers; Programme for Asbestos Abatement in Poland 2009–2032*; Polish Government: Warsaw, Poland, 15 March 2010.
11. *European Parliament resolution 2012/2065 (INI), Asbestos-Related Occupational health Threats and Prospects for Abolishing All Existing Asbestos*; European Parliament: Brussels, Belgium, 14 March 2013.

12. Li, J.; Dong, Q.; Yu, K.; Liu, L. Asbestos and asbestos waste management in the Asian-Pacific region: Trends, challenges and solutions. *J. Clean. Prod.* **2014**, *81*, 218–226. [CrossRef]

13. Paglietti, F.; Malinconico, S.; Conestabile della Staffa, B.; Bellagamba, S.; De Simone, P. Classification and management of asbestos-containing waste: European legislation and the Italian experience. *Waste Manag.* **2016**, *50*, 130–150. [CrossRef] [PubMed]

14. Asbestos Datebase. Available online: https://www.bazaazbestowa.gov.pl/en/ (accessed on 5 October 2018).

15. Gualtieri, A.F.; Tartaglia, A. Thermal decomposition of asbestos and recycling in traditional ceramics. *J. Eur. Ceram. Soc.* **2000**, *20*, 1409–1418. [CrossRef]

16. Gualtieri, A.F.; Cavenati, C.; Zanatto, I.; Meloni, M.; Elmi, G.; Gualtieri, M.L. The transformation sequence of cement–asbestos slates up to 1200 °C and safe recycling of the reaction product in stoneware tile mixtures. *J. Hazard. Mater.* **2008**, *152*, 563–570. [CrossRef] [PubMed]

17. Dellisanti, F.; Rossi, P.L.; Valdrè, G. Remediation of asbestos containing materials by Joule heating vitrification performed in a pre-pilot apparatus. *Int. J. Miner. Process.* **2009**, *91*, 61–67. [CrossRef]

18. Kusiorowski, R.; Zaremba, T.; Piotrowski, J.; Podwórny, J. Utilisation of cement-asbestos wastes by thermal treatment and the potential possibility use of obtained product for the clinker bricks manufacture. *J. Mater. Sci.* **2015**, *50*, 6757–6767. [CrossRef]

19. Leonelli, C.; Veronesi, P.; Boccaccini, D.N.; Rivasi, M.R.; Barbieri, L.; Andreola, F.; Lancellotti, I.; Rabitti, D.; Pellacani, G.C. Microwave thermal inertisation of asbestos containing waste and its recycling in traditional ceramics. *J. Hazard. Mater.* **2006**, *135*, 149–155. [CrossRef]

20. Anastasiadou, K.; Axiotis, D.; Gidarakos, E. Hydrothermal conversion of chrysotile asbestos using near supercritical conditions. *J. Hazard. Mater.* **2010**, *179*, 926–932. [CrossRef]

21. Viani, A.; Gualtieri, A.F.; Pollastri, S.; Rinaudo, C.; Croce, A.; Urso, G. Crystal chemistry of the high temperature product of transformation of cement-asbestos. *J. Hazard. Mater.* **2013**, *248*, 69–80. [CrossRef]

22. Gualtieri, A.F.; Boccaletti, M. Recycling of the product of thermal inertization of cement–asbestos for the production of concrete. *Constr. Build. Mater.* **2011**, *25*, 3561–3569. [CrossRef]

23. Viani, A.; Gualtieri, A.F. Preparation of magnesium phosphate cement by recycling the product of thermal transformation of asbestos containing wastes. *Cem. Concr. Res.* **2014**, *58*, 56–66. [CrossRef]

24. Gualtieri, A.F.; Giacobbe, C.; Sardisco, L.; Saraceno, M.; Gualtieri, M.L.; Lusvardi, G.; Cavenati, C.; Zanatto, I. Recycling of the product of thermal inertization of cement–asbestos for various industrial applications. *Waste Manag.* **2011**, *31*, 91–100. [CrossRef] [PubMed]

25. Viani, A.; Gualtieri, A.F. Recycling the product of thermal transformation of cement-asbestos for the preparation of calcium sulfoaluminate clinker. *J. Hazard. Mater.* **2013**, *260*, 813–818. [CrossRef] [PubMed]

26. Kusiorowski, R.; Zaremba, T.; Piotrowski, J. The potential use of cement-asbestos waste in the ceramic masses destined for sintered wall clay brick manufacture. *Ceram. Int.* **2014**, *40*, 11995–12002. [CrossRef]

27. Croce, A.; Allegrina, M.; Trivero, P.; Rinaudo, C.; Viani, A.; Pollastri, S.; Gualtieri, A.F. The concept of 'end of waste' and recycling of hazardous materials: In depth characterization of the product of thermal transformation of cement-asbestos. *Mineral. Mag.* **2014**, *78*, 1177–1191. [CrossRef]

28. Belardi, G.; Piga, L. Influence of calcium carbonate on the decomposition of asbestos contained in end-of-life products. *Thermochim. Acta* **2013**, *573*, 220–228. [CrossRef]

29. Yoshikawa, N.; Kashimura, K.; Hashiguchu, M.; Sato, M.; Horikoshi, S.; Mitani, T.; Shinohara, M. Detoxification mechanism of asbestos materials by microwave treatment. *J. Hazard. Mater.* **2015**, *284*, 201–206. [CrossRef]

30. Bloise, A.; Catalano, M.; Barrese, E.; Gualtieri, A.F.; Gandolfi, N.B.; Capella, S.; Belluso, E. TG/DSC study of the thermal behaviour of hazardous mineral fibers. *J. Therm. Anal. Calorim.* **2016**, *123*, 2225–2239. [CrossRef]

31. Yvon, Y.; Sharrock, P. Characterization of thermochemical inactivation of asbestos containing wastes and recycling the mineral residues in cement products. *Waste Biomass Valoriz.* **2011**, *2*, 169–181. [CrossRef]

32. Kusiorowski, R.; Zaremba, T.; Piotrowski, J.; Adamek, J. Thermal decomposition of different types of asbestos. *J. Therm. Anal. Calorim.* **2012**, *109*, 693–704. [CrossRef]

33. Kusiorowski, R.; Zaremba, T.; Piotrowski, J.; Gerle, A. Thermal decomposition of asbestos-containing materials. *J. Therm. Anal. Calorim.* **2013**, *113*, 179–188. [CrossRef]

34. Osada, M.; Takamiya, K.; Manako, K.; Noguchi, M.; Sakai, S. Demonstration study of high temperature melting for asbestos-containing waste (ACM). *J. Mater. Cycles Waste Manag.* **2013**, *15*, 25–36. [CrossRef]

35. Ruiz, A.I.; Ortega, A.; Fernández, R.; Miranda, J.F.; López Samaniego, E.; Cuevas, J. Thermal treatment of asbestos containing materials (ACM) by mixing with Na_2CO_3 and special clays for partial vitrification of waste. *Mater. Lett.* **2018**, *232*, 29–32. [CrossRef]

36. Averroes, A.; Sekiguchi, H.; Sakamoto, K. Treatment of airborne asbestos and asbestos-like microfiber particles using atmospheric microwave air plasma. *J. Hazard. Mater.* **2011**, *195*, 405–413. [CrossRef] [PubMed]

37. Iwaszko, J.; Zawada, A.; Przerada, I.; Lubas, M. Structural and microstructural aspects of asbestos-cement waste vitrification. *Spectrochim. Acta A Mol. Biomol. Spectrosc.* **2018**, *195*, 95–102. [CrossRef]

38. Sugama, T.; Sabatini, R.; Petrakis, L. Decomposition of chrysotile asbestos by fluorosulfonic acid. *Ind. Eng. Chem. Res.* **1998**, *37*, 79–88. [CrossRef]

39. Pawełczyk, A.; Bożek, F.; Grabas, K.; Chęcmanowski, J. Chemical elimination of the harmful properties of asbestos from military facilities. *Waste Manag.* **2017**, *61*, 377–385. [CrossRef]

40. Yanagisawa, K.; Kozawa, T.; Onda, A.; Kanazawa, M.; Shinohara, J.; Takanami, T.; Shiraishi, M. A novel decomposition technique of friable asbestos by $CHClF_2$-decomposed acidic gas. *J. Hazard. Mater.* **2009**, *163*, 593–599. [CrossRef]

41. Rozalen, M.; Huertas, F.J. Comparative effect of chrysotile leaching in nitric, sulfuric and oxalic acids at room temperature. *Chem. Geol.* **2013**, *352*, 134–142. [CrossRef]

42. Nam, S.N.; Jeong, S.; Lim, H. Thermochemical destruction of asbestos-containing roofing slate and the feasibility of using recycled waste sulfuric acid. *J. Hazard. Mater.* **2014**, *265*, 151–157. [CrossRef] [PubMed]

43. Turci, F.; Tomatis, M.; Mantegna, S.; Cravotto, G.; Fubini, B. A new approach to the decontamination of asbestos-polluted waters by treatment with oxalic acid under power ultrasound. *Ultrason. Sonochem.* **2008**, *15*, 420–427. [CrossRef] [PubMed]

44. Plescia, P.; Gizzi, D.; Benedetti, S.; Camilucci, L.; Fanizza, C.; De Simone, P.; Paglietti, F. Mechanochemical treatment to recycling asbestos-containing waste. *Waste Manag.* **2003**, *23*, 209–218. [CrossRef]

45. Colangelo, F.; Cioffi, R.; Lavorgna, M.; Verdolotti, L.; De Stefano, L. Treatment and recycling of asbestos-cement containing waste. *J. Hazrd. Mater.* **2011**, *195*, 391–397. [CrossRef] [PubMed]

46. Gualtieri, A.F. Recycling asbestos-containing material (ACM) from construction and demolition waste (CDW). In *Handbook of Recycled Concrete and Demolition Waste*; Pacheco-Torgal, F., Tam, V.W.Y., Labrincha, J.A., Ding, Y., de Brito, J., Eds.; Woodhead Publishing: Cambridge, UK, 2013; Volume 47, pp. 500–525, ISBN 978-0-85709-682-1. [CrossRef]

47. Spasiano, D.; Pirozzi, F. Treatments of asbestos containing wastes. *J. Environ. Manag.* **2017**, *204*, 82–91. [CrossRef] [PubMed]

48. Paolini, V.; Tomassetti, L.; Segreto, M.; Borin, D.; Liotta, F.; Torre, M.; Petracchini, F. Asbestos treatment technologies. *J. Mater. Cycles Waste Manag.* **2018**. [CrossRef]

49. Witek, J.; Kusiorowski, R. Neutralization of cement-asbestos waste by melting in an arc-resistance furnace. *Waste Manag.* **2017**, *69*, 336–345. [CrossRef]

50. Priyadharshini, P.; Mohan Ganesh, G.; Santhi, A.S. A review on artificial aggregates. *Int. J. Earth Sci. Eng.* **2012**, *5*, 540–546.

51. Tahmoorian, F.; Samali, B.; Tam, V.W.Y.; Yeaman, J. Evaluation of mechanical properties of recycled material for utilization in asphalt mixtures. *App. Sci.* **2017**, *7*, 763. [CrossRef]

52. *PN-EN ISO 12677:2011 Standard. Chemical Analysis of Refractory Products by X-ray Fluorescence (XRF)—Fused Cast-Bead Method*; ISO: Geneva, Switzerland, 2011.

53. *PN-EN 12620:2010 Standard; Aggregates for Concrete*; Polish Committee for Standardization: Warsaw, Poland, 2010.

54. *PN-EN 13043:2004 Standard; Aggregates for Bituminous Mixtures and Surface Treatments for Roads, Airfields and Other Trafficked Areas*; Polish Committee for Standardization: Warsaw, Poland, 2010.

55. *PN-EN 933-1:2012 Standard; Tests for Geometrical Properties of Aggregates—Part 1: Determination of Particle Size Distribution—Sieving Method*; Polish Committee for Standardization: Warsaw, Poland, 2012.

56. *PN-EN 933-3:2012 Standard; Tests for Geometrical Properties of Aggregates—Part 3: Determination of Particle Shape—Flakiness Index*; Polish Committee for Standardization: Warsaw, Poland, 2012.

57. *PN-EN 933-5:2000 Standard; Tests for Geometrical Properties of Aggregates—Part 5: Determination of the Percentage of Crushed and Broken Surfaces in Coarse Aggregate Particles*; Polish Committee for Standardization: Warsaw, Poland, 2000.

58. *PN-EN 1097-6:2013 Standard; Tests for Mechanical and Physical Properties of Aggregates—Part 6: Determination of Particle Density and Water Absorption*; Polish Committee for Standardization: Warsaw, Poland, 2013.
59. *PN-EN 1097-2 2010 Standard; Tests for Mechanical and Physical Properties of Aggregates—Part 2: Methods for the Determination of Resistance to Fragmentation*; Polish Committee for Standardization: Warsaw, Poland, 2010.
60. *PN-EN 1367-1 2007 Standard; Tests for Thermal and Weathering Properties of Aggregates—Part 1: Determination of Resistance to Freezing and Thawing*; Polish Committee for Standardization: Warsaw, Poland, 2007.
61. Viani, A.; Gualtieri, A.F.; Secco, M.; Peruzzo, L.; Artioli, G.; Cruciani, G. Crystal chemistry of cement-asbestos. *Am. Mineral.* **2013**, *98*, 1095–1105. [CrossRef]
62. Stepkowska, E.T.; Blanes, J.M.; Franco, F.; Real, C.; Perez-Rodriguez, J.L. Phase transformation on heating of an aged cement paste. *Thermochim. Acta* **2004**, *420*, 79–87. [CrossRef]
63. Zulumyan, N.; Mirgorodski, A.; Isahakyan, A.; Beglaryan, H. The mechanism of decomposition of serpentines from peridotites on heating. *J. Therm. Anal. Calorim.* **2014**, *115*, 1003–1012. [CrossRef]

fibers

MDPI

Article

Assessment of Serpentine Group Minerals in Soils: A Case Study from the Village of San Severino Lucano (Basilicata, Southern Italy)

Rosalda Punturo [1], Claudia Ricchiuti [1] and Andrea Bloise [2,*]

[1] Department of Biological, Geological and Environmental Sciences, University of Catania, Corso Italia, 55, 95129 Catania, CT, Italy; punturo@unict.it (R.P.); claudia.ricchiuti@unict.it (C.R.)
[2] Department of Biology, Ecology and Earth Sciences, University of Calabria, Via Pietro Bucci, I-87036 Rende, Italy
*   Correspondence: andrea.bloise@unical.it; Tel.: +39-0984-493588

Received: 30 January 2019; Accepted: 19 February 2019; Published: 25 February 2019

Abstract: Naturally occurring asbestos (NOA) is a generic term used to refer to both regulated and un-regulated fibrous minerals when encountered in natural geological deposits. These minerals represent a cause of health hazard, since they have been assessed as potential environmental pollutants that may occur both in rocks and derived soils. In the present work, we focused on the village of San Severino Lucano, located in the Basilicata region (southern Apennines); due to its geographic isolation from other main sources of asbestos, it represents an excellent example of hazardous and not occupational exposure of population. From the village and its surroundings, we collected eight serpentinite-derived soil samples and carried out Differential Scanning Calorimetry (DSC), Derivative Thermogravimetric (DTG) and Transmission Electron Microscopy with Energy Dispersive Spectrometry (TEM-EDS), in order to perform a detailed characterization of serpentine varieties and other fibrous minerals. Investigation pointed out that chrysotile and asbestos tremolite occur in all of the samples. As for the fibrous but non-asbestos classified minerals, polygonal serpentine and fibrous antigorite were detected in a few samples. Results showed that the cultivation of soils developed upon serpentinite bedrocks were rich in harmful minerals, which if dispersed in the air can be a source of environmental pollution.

Keywords: serpentine varieties; naturally occurring asbestos; health hazard; serpentinite soil

---

1. Introduction

As it is known, the term "asbestos" represents a group of six fibrous silicate minerals: chrysotile (serpentine group) and amphibole group as: tremolite, actinolite, anthophyllite, amosite and crocidolite [1,2]. In the past, asbestos was plenty been exploited and marketed for the use in industrial and commercial products, mainly as building material [3]. All types of asbestos cause lung cancer, mesothelioma, cancer of the larynx and ovary, and asbestosis (fibrosis of the lungs) [2]. It has been assessed that exposure to asbestos occurs through inhalation of airborne fibers in various contexts such as the working environment, ambient air in the vicinity of point sources such as factories handling asbestos, or indoor air in housing and buildings containing friable asbestos materials [4]. Nevertheless, it is worth noting that natural occurrences of asbestos represent a cause of health hazard, which is sometimes overlooked and difficult to properly monitor. Indeed, naturally occurring asbestos (NOA) is a generic term used to refer to both regulated and non-regulated fibrous minerals when encountered in natural geological deposits [5]. Now-a-days, only the six varieties above listed are regulated as potential environmental pollutants by law (in Europe and in several countries worldwide), even though other asbestiform minerals such as balangeorite, erionite, fibrous antigorite and fluoro-edenite [6–9] are non-asbestos classified and, therefore, not regulated by law but could be potentially dangerous if inhaled. On the basis of the effects of asbestos on biological systems, several authors ascribe the asbestos-fibers toxicity to the synergetic effect of fiber size, bio-persistence and chemical composition [10–13]; this latter is related to the high capability of asbestos minerals to host a large number of toxic elements; for this reason, due to interactions between lung fluids and inhaled atmospheric dust [14,15], some researchers claimed that asbestos fibers may play a passive role in producing diseases as carriers of heavy metals that may be then released into the environment [16]. In general, many factors such as natural weathering processes (e.g., erosion) and human activities (e.g., excavation, road construction, agricultural activities) contribute to NOA release in the environment [13,17], enhancing hazard of people who live near to NOA deposits around the world [18–25].

In the present study we focused on the Basilicata region (Italy) [26], where an increased number of lung disease cases were related to the environmental exposure to asbestos [27–29]. The village and its surroundings represent an excellent example of hazardous and not occupational exposure of population to asbestos, because of the geographic isolation and its distance from other main sources of asbestos for instance.

Recently, a work by Punturo et al. [30] dealt with the characterization by X-Ray Fluorescence (XRF), X-Ray Powder Diffraction (XRPD) and Scanning and Electron Microscopy (SEM) of the soils of San Severino Lucano, reporting their potential for hazardous exposure of population, because of their heavy metal content. However, the discrimination among the serpentine group minerals (i.e., lizardite, antigorite chrysotile, polygonal serpentine) was not achievable by using only X-ray powder diffraction, because the diffraction peaks overlap each other. Moreover, scanning electron microscopy (SEM) alone could not determine the diameter of single fibrils. Since these last techniques were not able to identify the different serpentine varieties, in this work a more targeted characterization of soil samples was performed by Differential Scanning Calorimetry (DSC), Derivative thermogravimetric (DTG) and Transmission Electron Microscopy with Energy Dispersive Spectrometry (TEM-EDS). We collected eight serpentinite soil samples and cross-checked the data obtained from DSC, DTG and TEM-EDS, in order to perform a detailed characterization and discrimination among the serpentine varieties and other fibrous minerals, as well as to relate NOA release in the environment due to agricultural activity. Investigation highlighted that chrysotile and asbestos tremolite are the asbestos minerals occurring in all of the analyzed soils, appearing both as single fibrils and bundles. As for the fibrous but non-asbestos classified minerals, polygonal serpentine and fibrous antigorite were detected in a few samples. Because of the fibrous structure, longitudinal splitting of these minerals is very common, creating thus fibers having the same length as the original one but with smaller diameter. Furthermore, the cultivation of soils developed on serpentinite bedrocks could enhance this process and provoke the release of smaller fibrils into the environment, increasing thus the exposure of population to asbestos risk. Results may provide a useful tool for planning prevention measures during agricultural activities, in order to diminish negative effects of NOA on health.

2. Geological Setting

This study area encompasses approximately 20 km^2 in the Pollino National Park [16], which is located at the borders between the Basilicata and Calabria regions (southern Italy; Figure 1).

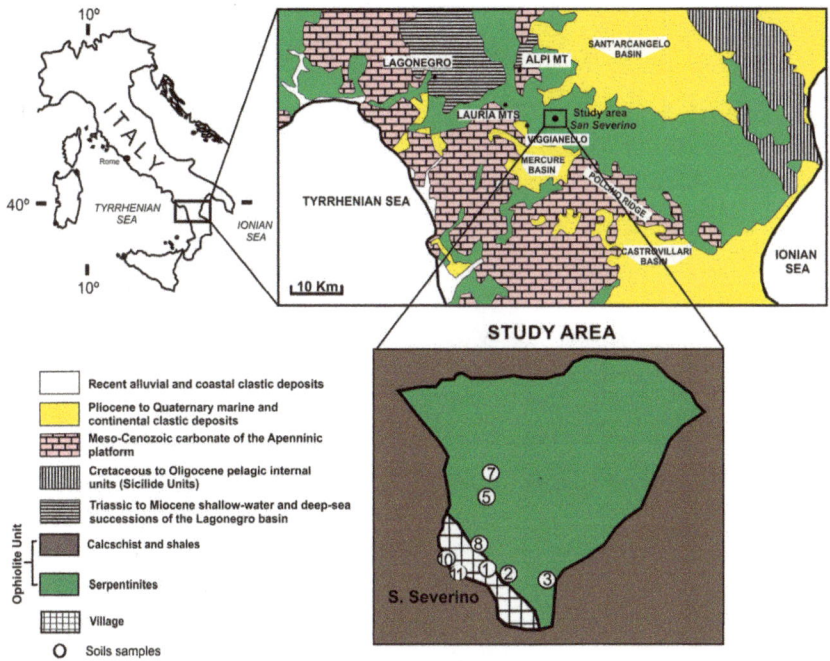

Figure 1. Geological map of the Calabria-Lucania border (modified after [30]) and study area location with sampling sites.

The area is characterized by the terrains of the Liguride Complex, which consists of three main tectonic units of Upper Jurassic to Upper Oligocene age [31]: (1) the Calabro-Lucano Flysch [32], a unit that did not underwent any metamorphism and partly corresponds to the North-Calabrian Unit; (2) the metamorphic terranes of the Frido Unit [33,34]; (3) syn-orogenic turbiditic sequences, i.e., the Saraceno Formation, the Albidona Formation, and a sequence composed of alternating shales, mudstones and sandstones, the latter corresponding to the Perosa Unit as defined by Vezzani [35]. The ophiolites of the Southern Apennine Liguride Units occur in the Frido Unit and in the North-Calabrian Unit. In particular, ophiolitic rocks of the Frido Unit consist of lenticular metabasites interbedded with cataclastic and highly fractured serpentinite rocks [36]. Metabasite rocks are foliated and fine-grained, with rare remnants of porphyritic texture. They are often intercalated with serpentinites, slates and metacarbonate rocks [37], forming sequences with a maximum thickness of several dozen metres. Serpentinite rocks, which are green-bluish in colour, represent mantle peridotites [38]. Locally, serpentinites are very brittle, as indicated by the large number of fractures that are usually filled by amphibole asbestos. As it may be observed on Figure 1, serpentinite lithotypes constitute the bedrock of the village of San Severino Lucano and its surroundings. The detailed field survey carried out along the transect of sampling sites located at San Severino village and its surroundings (Figure 1), highlighted that the area is characterized by sparse vegetation and by soils developed on serpentinite bedrocks (Figure 2a–e).

Figure 2. (**a**) Distant view of the San Severino Village (modified after [26]); (**b**) soil outcrop that contain NOA (Spol8); (**c**) soil outcrop (Spol1); (**d**) soil outcrop (Spol10); (**e**) soil outcrop (Spol2).

3. Materials and Methods

Eight serpentinite derivative soil samples (Spol1,2,3,5,7,8,10,11) were collected mainly within to urban center and analyzed by using TEM-EDS and thermal analyzes (DSC, DTG) at the University of Calabria (DiBEST laboratory), in order to investigate their mineralogical features and to assess the occurrence of asbestiform minerals, which are considered to be potentially hazardous for human health [13]. It is worth mentioning that combination of both analytical methodologies, i.e., thermal analysis and TEM-EDS, permitted successful identification of distinct serpentine minerals (antigorite, lizardite, chrysotile, and polygonal serpentine) and the characterization of amphibole asbestos [13]. Moreover, the length of fibrous antigorite and polygonal serpentine fibers has been measured using the TEM micrographs, adding further details to the previous observations carried out with SEM [30]. The soil samples were pre-treated with H_2O_2 and pre-heated for 24 h at 530 °C, in order to remove the organic compounds and so that they could be subsequently ground. Size and chemical composition of single fibers were determined using a Jeol JEM 1400 Plus (120 kV) Transmission Electron Microscope equipped with Jeol large-area silicon drift detector SDD-EDS (Jeol, Tokyo, Japan) for microanalyses. For TEM investigation, each sample was put into isopropyl alcohol and then sonicated. Three drops of the obtained suspension were deposited on a Formvar carbon-coated copper grid.

Thermogravimetry (TG) and differential scanning calorimetry (DSC) were performed in an alumina crucible under a constant aseptic air flow of 30 mL·min^{-1} with a Netzsch STA 449 C

Jupiter (Netzsch-Gerätebau GmbH, Selb, Germany) in the 25–1000 °C temperature range with a heating rate of 10 °C·min^{-1} and 20 mg of sample powder. Instrumental precision was checked by 5 repeated collections on a kaolinite reference sample revealing good reproducibility (instrumental theoretical T precision of ±1.2 °C). Netzsch Proteus thermal analysis software (Netzsch-Gerätebau GmbH, Selb, Germany) was used to identify exo- and endothermic peaks, weight loss and derivative thermogravimetric (DTG).

4. Results

4.1. TEM Characterization

TEM has been mainly useful to determine the occurrence of serpentine varieties and their morphological features in the soil samples; indeed, distinct fibrous serpentine varieties have been found such as chrysotile, fibrous antigorite, polygonal serpentine and tremolite (Figure 3). They exhibit various shape and size. Chrysotile appears as thin individual fibers (known as fibrils) and often forms relatively larger longitudinally aligned fibers (Figure 3a,b).

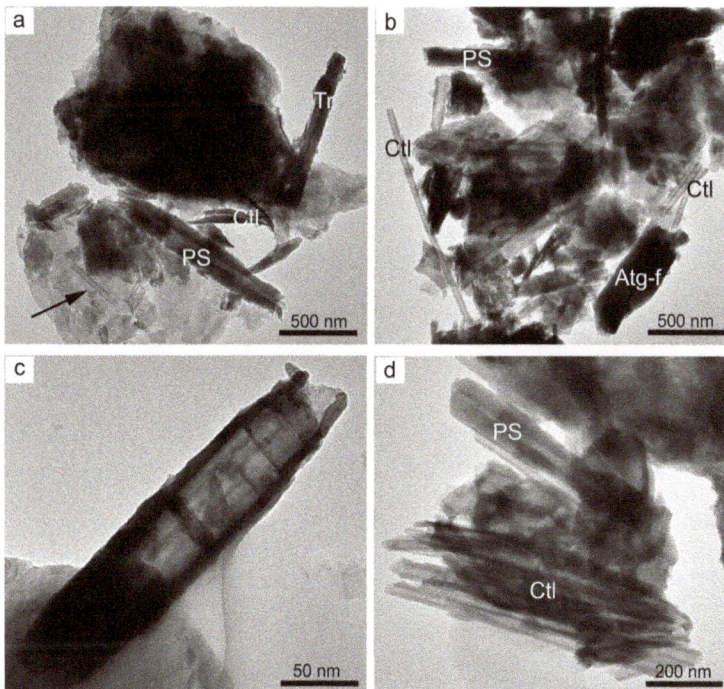

Figure 3. Representative TEM images of fibrous mineral detected in the soil samples: (**a**) bundles of chrysotile fibers; chrysotile fiber characterized by the empty central cavity and thin outer walls indicated by black arrow; polygonal serpentine and fibrous tremolite (sample Spol1); (**b**) chrysotile fibers, polygonal serpentine and fibrous antigorite (Spol3); (**c**) chrysotile fiber partially unrolled from the inside like cylinder-in-cylinder morphology (Spol10); (**d**) polygonal serpentine and bundles of chrysotile fibers (Spol11). Ctl = chrysotile; Tr = tremolite; Atg-f = fibrous antigorite; PS = polygonal serpentine (mineral symbols after Whitney and Evans [39]).

From Figures 3 and 4 it is evident the classical cylindrical shape of chrysotile fibers; this is the most common morphology in all of the samples, consisting of an empty central cavity (core) along throughout their length. The length varies from 300 to 1500 nm and the diameter of the core is about

20 nm and 40 nm inner and outer, respectively. In some samples the outer walls of chrysotile are very thin and the central tube (core) is wide, measuring about 40 nm (Figure 3a,b). This proves that the chrysotile underwent an unrolling process from the inside during the process of alteration from rock to soil, likely caused by the passage of water through the core (Figure 3c). Chrysotile with cylinder-en-cylinder and proto-cylinder morphologies have also been found with TEM investigation; these do not show the well-defined wrapping of layers and cylindrical shape that the chrysotile fibers exhibited (Figure 4).

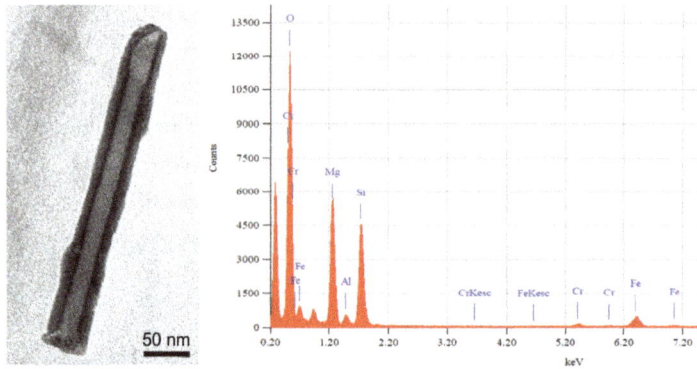

Figure 4. Single cylinder chrysotile with the relative point analysis (Spol1).

Fibrous polygonal serpentine is another structural variety that has been found in most of the studied samples (Table 1); it occurs in lower amount and has very often a diameter larger than 100 nm and wider than the chrysotile individuals (Figure 3a,b,d). Antigorite fibers are the shortest, with length and width of 1000 and 300 nm respectively. However, fibrous antigorite has been identified only in two samples (Figure 3b; Table 1), with platy antigorite the most abundant morphology observed in all of the studied specimens. Lizardite with platy morphology was also detected in a few samples (Table 1). Tremolite fibers have also been observed. TEM micrographs reported on Figures 3a and 5, show the typical morphology of tremolite fibers, which exhibit prismatic rod-shaped morphology lacking of any flexibility. In these fibers, the average length ranges from 2.5 μm to 3 μm and the diameter is about 0.2 μm.

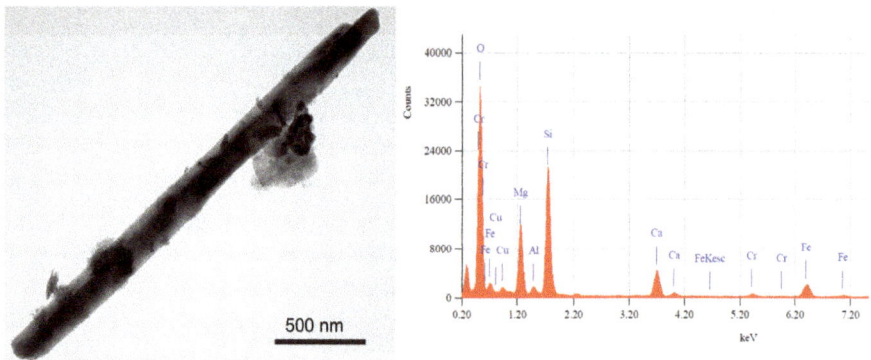

Figure 5. Single tremolite fiber with the relative point analysis (Spol1).

Table 1. Studied localities, reference coordinates and, for each collected soil sample, mineralogical assemblage detected by X-ray powder diffraction (XRPD) and by scanning electron microscopy combined with energy dispersive spectrometry (SEM-EDS) * after [30]. Serpentine minerals varieties and amphiboles detected by DSC, DTG and TEM-EDS. Chlorite (Chl), chrysotile (Ctl), polygonal serpentine (PS), lizardite (Liz), fibrous antigorite (f-Ant), antigorite (Atg) and tremolite (Tr) (mineral symbols after Whitney and Evans [39]). Amphiboles present in the samples were classified according to the amphibole diagram classification [40].

Sample	Site Description	Longitude (East)	Latitude (North)	Phases Detected
Spol1	At the entrance of the Village	597,417	4,429,775	Ctl, PS, Ant, Tr (Di, Qtz, Mnt-Chl) *
Spol2	At the entrance of the Village	597,405	4,430,523	Ctl, f-Ant (Di, Qtz, Mnt-Chl, Tr) *
Spol3	Road cut outside the Village	597,808	4,430,474	Ctl, PS, Liz, f-Ant (Di, Qtz, Mnt-Chl, Tr) *
Spol5	Road cut outside the Village	597,270	4,431,103	Ctl, Liz, Ant (Di, Qtz, Mnt-Chl, Tr, Chm, Ms) *
Spol7	Road cut outside the Village	597,323	4,431,363	Ctl, Tr, (Di, Qtz, Mnt-Chl, Chm) *
Spol8	Road cut within the Village	597,223	4,430,711	Ctl, PS, Ant (Di, Qtz, Mnt-Chl, Tr, Chm) *
Spol10	Road cut within the Village	596,890	4,430,715	Ctl, PS, Ant (Di, Qtz, Mnt-Chl, Tr, Chm, Mo) *
Spol11	Road cut within the Village	596,890	4,430,715	Ctl, PS (Di, Qtz, Mnt, Tr, Chm) *

4.2. Thermal Analysis Characterization

Thermal analysis of all representative soil samples enabled us to recognize the constituent mineralogical phases, and in particular the serpentine varieties (i.e., antigorite, lizardite, chrysotile, polygonal serpentine) (Table 1). In Figure 6a, the DSC patterns describe the thermal behavior of the investigated samples. In the temperature range between 500 and 850 °C, chrysotile lost its chemical-bonded water (strong endothermic peak on average temperature at 630 °C, Figure 6b) causing the complete breakdown of the mineral structure. At higher temperature value, the crystallization of forsterite [41] generates a sharp exothermic peak recorded at about 830 °C (Figure 6a; Table 2). After thermal analysis, the chrysotile structure has completely changed at a molecular scale because of a phenomenon called pseudomorphosis, which leads to the complete transformation of asbestos minerals into non-hazardous silicates such as forsterite [42,43].

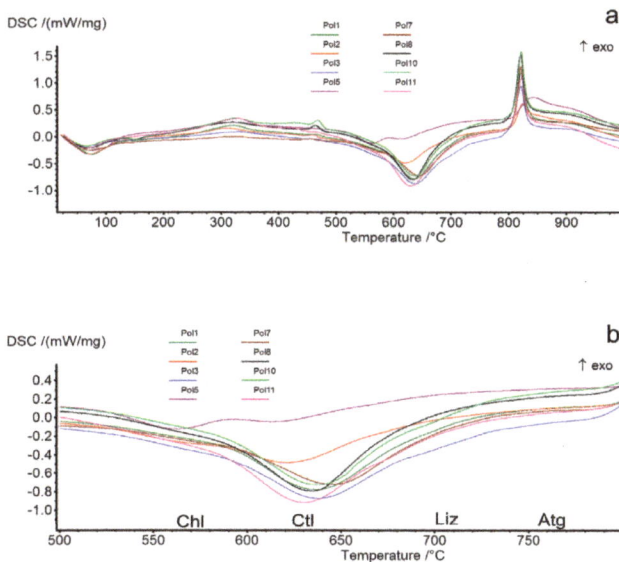

Figure 6. (**a**) Comparison among differential scanning calorimetry (DSC) curves of the soils located within or near the village of San Severino Lucano Village (Basilicata, Southern Italy); (**b**) zoom of (**a**) in the temperature range of 500–800 °C.

Table 2. Peak temperatures in DSC and DTG curves. W = weak, vw = very weak, s = strong, ss = very strong, sh = shoulder, en = endothermic, ex = exothermic.

Samples	Spol1	Spol2	Spol3	Spol5	Spol7	Spol8	Spol10	Spol11
				DSC				
Chl				563 en(w)				
Ctl	636 en(s)	621 en(s)	638 en(s)	612 en(w)	645 en(w)	634 en(s)	637 en(s)	630 en(s)
Fo	822 ex(ss)	824 ex(ss)	821 ex(ss)	844 ex(s)	821 ex(ss)	821 ex(ss)	822 ex(ss)	822 ex(ss)
				DTG				
Chl		564 en(vw)		563 en(w)				
Ctl	637 en(ss)	619 en(s)	638 en(s)	614 en(w)	647 en(ss)	634 en(s)	639 en(ss)	631 en(ss)
PS	679 en(vw)		686 en(vw)			677 en(vw)	688 en(vw)	679 en(vw)
Liz			736 en(sh)	744 en(sh)				
Ant	774 en(sh)	784 en(vw)	784 en(sh)	790 en(w)		778 en(sh)	770 en(sh)	

DTG curves appear to be similar for most of the samples and show the main endothermic peaks related to the mineralogical phases decomposition between 500 and 830 °C (Figure 7). A weak endothermic peak at 563 and 564 °C for Spol2 and Spol5 respectively, is linked to the presence of a small amount of chlorite. For all samples, the very strong endothermic peak in a temperature range of 614–639 °C, clearly showed the presence of chrysotile in high amount (Table 2).

Polygonal serpentine occurs in most of the samples showing a weak endothermic peak in a range of 677–688 °C, whereas only two samples, Spol3 and Spol5, are characterized by an endothermic shoulder at 736 and 744 °C related to the presence of lizardite (Table 2). Finally, the occurrence of antigorite is confirmed by the endothermic peak in a T range of 770–790 °C (Table 2).

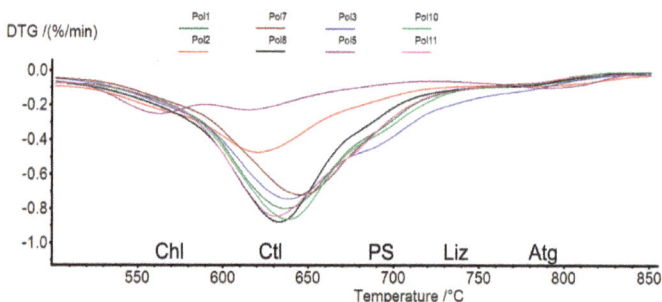

Figure 7. Comparison among DTG curves in the temperature range of 500–850 °C: endothermic peaks related to chlorite (Chl), chrysotile (Ctl), polygonal serpentine (PS), lizardite (Liz) and antigorite (Atg) decomposition.

The TG data reported on Table 3, show the values of 1–4% mass loss at a temperature up to 110 °C due to the adsorbed water and total weight losses of about 12–18% up to 1000 °C in all of the samples, mainly due to the breakdown of serpentine minerals according to the literature data [44,45]. Some samples have high values of total water loss at 1000 °C (i.e., 18%) due to two reasons: (i) presence of other hydrated minerals in addition to the serpentine polymorphs (Table 1); and (ii) the presence of water (physically bound) trapped between the fibrous bundles of chrysotile [46].

Table 3. TG data (weight loss % up to 110 °C and up to 1000 °C) for the analyzed samples.

Spol1		Spol2	
T range (°C)	TG loss %	T range (°C)	TG loss %
<110	3.57	<110	3.30
TOT loss at 1000	17.44	TOT loss at 1000	14.80
Spol3		**Spol5**	
T range (°C)	TG loss %	T range (°C)	TG loss %
<110	1.47	<110	3.94
TOT loss at 1000	13.13	TOT loss at 1000	12.08
Spol7		**Spol8**	
T range (°C)	TG loss %	T range (°C)	TG loss %
<110	2.16	<110	2.39
TOT loss at 1000	12.24	TOT loss at 1000	14.90
Spol10		**Spol11**	
T range (°C)	TG loss %	T range (°C)	TG loss %
<110	1.96	<110	2.78
TOT loss at 1000	15.40	TOT loss at 1000	15.42

5. Discussion and Conclusions

The results obtained by thermal analysis and transmission electron microscope showed that chrysotile and asbestos tremolite are the asbestos minerals occurring in all of the analyzed soil samples in the area of san Severino Lucano village (southern Apennines). Chrysotile appears both as bundles and single fibrils with typical cylindrical shape with diameter and length shorter of 0.25 and 5 µm, respectively; it is interesting to point out as the most common morphology of chrysotile fibers is the classical cylindrical shape consisting of an empty central cavity throughout the length. Moreover, some chrysotile fibers are characterized by very thin outer walls and by wide central tube, proving that chrysotile, during the process of alteration in the passage from rock to soils underwent unrolling process from the inside; this is likely caused by the passage of water through the core. As far as occurring amphibole, results showed that it is tremolite, whose fibers exhibit prismatic rod-shaped morphology lacking of any flexibility. The average length ranges from 2.5 µm to 3 µm and the diameter is about 0.2 µm.

According to many authors, fibers shorter than 5 µm and very thin <0.25 µm may have considerable carcinogenic potential [47–49]. Therefore, both techniques applied revealed to be a useful tool for determining the occurrence of asbestiform varieties and their morphological features in the studied soil samples, permitting asbestos minerals to be univocally identified and investigated in detailed, revealing that the fibers found within the studied soil samples show a size that may be associated with carcinogenesis when breathed.

It is important to specify that, in addition to the minerals regulated as asbestos by the Italian law, also asbestiform minerals such as fibrous antigorite could be potentially dangerous if inhaled [9]. In this work, polygonal serpentine and antigorite were the fibrous minerals detected in five and two samples respectively, while the other minerals identified were non-fibrous and most of them showing platy morphology (e.g., lizardite, chlorite). The village of San Severino is a significant example of a settlement built on NOA-bearing outcrops [50,51] and the risk of inhaling airborne fibers of asbestos around the village increases due to the agricultural activities which are among the main resources for the economy of the area. The cultivation of soils developed on serpentinite bedrocks could provoke the fiber splitting into smaller fibrils that are widely spread out into the environment, increasing thus the exposure to them. Since asbestos occurrence in soils is a serious health problem, in many parts of the world, asbestos-containing land has been abandoned and countries with this problem have suffered economic losses due to depreciation of properties. The use of soils containing asbestos for agricultural purposes can increase the presence of fibers in the air, necessitating adequate attention to ensuring the protection of workers and general public, as already pointed out by dedicated agencies. It is useless to

Fibers **2019**, *7*, 18

create unjustified alarmism in population; at the same time, inhabitants who live in countryside areas where NOA is present, should be aware that as now-a-days various techniques are available to limit or eliminate the presence of airborne fibers deriving from the processing of the soil, thus diminishing the risks related; among them, it is worth mentioning: (i) the use of tractors with air-conditioned and filtered cabins; (ii) wet the ground before hoeing it; (iii) wear overalls and masks suitable for protection from airborne asbestos fibers.

In conclusion, results presented in this work may provide a useful tool for planning prevention measures during human activities, in order to diminish negative effects of NOA on health.

Author Contributions: Conceptualization, A.B. and R.P.; Methodology (DSC-DTG and TEM-EDS measurements), A.B.; Software, A.B.; Validation, A.B., R.P. and C.R.; Formal Analysis, A.B.; Investigation, A.B., R.P. and C.R.; Resources, A.B. and R.P.; Data Curation, A.B. and R.P.; Writing-Original Draft Preparation, A.B., R.P. and C.R.; Writing-Review & Editing, A.B., R.P. and C.R.; Visualization, A.B.; Supervision, R.P. and A.B.; Project Administration, R.P.; Funding Acquisition, R.P. and A.B.

Funding: Part of this research was carried out under the financial support of "Piano Triennale della Ricerca (2017–2020)" (Università di Catania, Dipartimento di Scienze Biologiche, Geologiche e Ambientali), scientific responsible Rosalda Punturo. The work has received financial support from the FFABR fund (by the Italian MIUR) scientific responsible Andrea Bloise.

Acknowledgments: The authors thank E. Barrese for the support during data collection. The work has received financial support from University of Calabria and University of Catania.

Conflicts of Interest: The authors declare no conflict of interest.

References

1. World Health Organization (WHO). *Asbestos and Other Natural Mineral Fibers. Environmental Health Criteria, 53*; World Health Organization: Geneva, Switzerland, 1986; p. 194.
2. National Institute for Occupational Safety and Health (NIOSH). *Asbestos and Other Elongated Mineral Particles: State of the Science and Roadmap for Research*; Current Intelligence Bulletin, June 2008-Revised Draft; National Institute for Occupational Safety and Health (NIOSH): Washington, DC, USA, 2008.
3. Punturo, R.; Cirrincione, R.; Pappalardo, G.; Mineo, S.; Fazio, E.; Bloise, A. Preliminary laboratory characterization of serpentinite rocks from Calabria (southern Italy) employed as stone material. *J. Mediterr. Earth Sci.* **2018**, *10*, 79–87.
4. International Agency for Research on Cancer (IARC). *Asbestos (Chrysotile, Amosite, Crocidolite, Tremolite, Actinolite, and Anthophyllite) IARC Monographs. Arsenic, Metals, Fibers and Dusts*; International Agency for Research on Cancer: Lyon, France, 2009; pp. 147–167.
5. Harper, M. 10th Anniversary critical review: naturally occurring asbestos. *J. Environ. Monit.* **2008**, *10*, 1394–1408. [CrossRef] [PubMed]
6. Compagnoni, R.; Ferraris, G.; Fiora, L. Balangeorite, a new fibrous silicate related to gageite from Balangero, Italy. *Am. Mineral.* **1983**, *68*, 214–219.
7. Ballirano, P.; Pacella, A.; Bloise, A.; Giordani, M.; Mattioli, M. Thermal Stability of Woolly Erionite-K and Considerations about the Heat-Induced Behaviour of the Erionite Group. *Minerals* **2018**, *8*, 28. [CrossRef]
8. Gianfagna, A.; Ballirano, P.; Bellatreccia, F.; Bruni, B.; Paoletti, E.; Oberti, R. Characterization of amphibole fibers linked to mesothelioma in the area of Biancavilla, eastern Sicily, Italy. *Mineralog. Mag.* **2003**, *67*, 1221–1229. [CrossRef]
9. Cardile, V.; Lombardo, L.; Belluso, E.; Panico, A.; Capella, S.; Balazy, M. Toxicity and Carcinogenicity Mechanisms of Fibrous Antigorite. *Int. J. Environ. Res. Public Health* **2007**, *4*, 1–9. [CrossRef] [PubMed]
10. Gualtieri, A.F. *Mineral Fibers: Crystalchemistry, Chemical-Physicalproperties, Biological Interaction and Toxicity*; European Mineralogical Union and Mineralogical Society of Great Britain and Ireland: London, UK, 2017; p. 533.
11. Pugnaloni, A.; Giantomassi, F.; Lucarini, G.; Capella, S.; Bloise, A.; Di Primio, R.; Belluso, E. Cytotoxicityinduced by exposure to natural and synthetic tremolite asbestos: An in vitro pilot study. *Acta Histochem.* **2013**, *115*, 100–112. [CrossRef] [PubMed]
12. Bloise, A.; Catalano, M.; Barrese, E.; Gualtieri, A.F.; Gandolfi, N.B.; Capella, S.; Belluso, E. TG/DSC study of the thermal behaviour of hazardous mineral fibers. *J. Therm. Anal. Calorim.* **2016**, *123*, 2225–2239. [CrossRef]

13. Bloise, A.; Punturo, R.; Catalano, M.; Miriello, D.; Cirrincione, R. Naturally occurring asbestos (NOA) in rock and soil and relation with human activities: The monitoring example of selected sites in Calabria (southern Italy). *Ital. J. Geosci.* **2016**, *135*, 268–279. [CrossRef]

14. Censi, P.; Zuddas, P.; Randazzo, L.A.; Tamburo, E.; Speziale, S.; Cuttitta, A.; Punturo, R.; Santagata, R. Source and nature of inhaled atmospheric dust from trace element analyses of human bronchial fluids. *Environ. Sci. Technol.* **2011**, *45*, 6262–6267. [CrossRef] [PubMed]

15. Censi, P.; Tamburo, E.; Speziale, S.; Zuddas, P.; Randazzo, L.A.; Punturo, R.; Aricò, P. Yttrium and lanthanides in human lung fluids, probing the exposure to atmospheric fallout. *J. Hazard. Mater.* **2011**, *186*, 1103–1110. [CrossRef] [PubMed]

16. Bloise, A.; Barca, D.; Gualtieri, A.F.; Pollastri, S.; Belluso, E. Trace elements in hazardous mineral fibres. *Environ. Pollut.* **2016**, *216*, 314–323. [CrossRef] [PubMed]

17. Punturo, R.; Bloise, A.; Critelli, T.; Catalano, M.; Fazio, E.; Apollaro, C. Environmental implications related to natural asbestos occurrences in the ophiolites of the Gimigliano-Mount Reventino Unit (Calabria, southern Italy). *Int. J. Environ. Res.* **2015**, *9*, 405–418.

18. Acosta, A.; Pereira, M.D.; Shaw, D.M.; Bea, F. Serpentinización de la peridotita de Ronda (cordillera Betica) comorespuesta a la interacción con fluidosricos en volátiles: comportamiento del boro. *Rev. Soc. Geol. Esp.* **1997**, *10*, 99–106.

19. Burragato, F.; Comba, P.; Baiocchi, V.; Palladino, D.M.; Simei, S.; Gianfagna, A.; Pasetto, R. Geo-volcanological, mineralogical and environmental aspects of quarry materials related to pleural neoplasm in the area of Biancavilla, Mount Etna (Eastern Sicily, Italy). *Environ. Geol.* **2005**, *47*, 855–868. [CrossRef]

20. Constantopoulos, S.H. Environmental mesothelioma associated with tremolite asbestos: Lessons from the experiences of Turkey, Greece, Corsica, New Caledonia and Cyprus. *Regul. Toxicol. Pharmacol.* **2008**, *52*, 110–115. [CrossRef] [PubMed]

21. Pereira, M.D.; Peinado, M.; Blanco, J.A.; Yenes, M. Geochemical characterization of serpentinites at cabo ortegal, northwestern Spain. *Can. Mineral.* **2008**, *46*, 317–327. [CrossRef]

22. Bloise, A.; Belluso, E.; Critelli, T.; Catalano, M.; Apollaro, C.; Miriello, D.; Barrese, E. Amphibole asbestos and other fibrous minerals in the meta-basalt of the Gimigliano-Mount Reventino Unit (Calabria, south-Italy). *Rend Online Soc Geol It.* **2012**, *21*, 847–848.

23. Navarro, R.; Pereira, D.; Gimeno, A.; Barrio, S.D. Verde Macael: A Serpentinite Wrongly Referred to as a Marble. *Geosciences* **2013**, *3*, 102–113. [CrossRef]

24. Gaggero, L.; Sanguineti, E.; Yus González, A.; Militello, G.M.; Scuderi, A.; Parisi, G. Airborne asbestos fibers monitoring in tunnel excavation. *J. Environ. Manag.* **2017**, *196*, 583–593. [CrossRef] [PubMed]

25. Worliczek, E. Naturally occurring asbestos: The perception of rocks in the mountains of New Caledonia. In *Environmental Transformations and Cultural Responses: Ontologies, Discourses, and Practices in Oceania*; Dürr, E., Pascht, A., Eds.; Springer: Berlin, Germany, 2017; pp. 187–214.

26. Bloise, A.; Catalano, M.; Critelli, T.; Apollaro, C.; Miriello, D. Naturally occurring asbestos: Potential for human exposure, San Severino Lucano (Basilicata, Southern Italy). *Environ. Earth Sci.* **2017**, *76*, 648. [CrossRef]

27. Bernardini, P.; Schettino, B.; Sperduto, B.; Giannadrea, F.; Burragato, F.; Castellino, N. Tre Casi di mesotelioma pleurico ed inquinamento ambientale da rocce affioranti di tremolite in Lucania. *GIMLE* **2003**, *25*, 408–411.

28. Burragato, F.; Mastacchi, R.; Papacchini, L.; Rossini, F.; Sperduto, B. Mapping of risks due to particulates of natural origin containing fibrous tremolite: The case of Seluci di Lauria (Basilicata, Italy). In Proceedings of the 1st General Assembly, Nice, France, 25–30 April 2004.

29. Pasetto, R.; Bruni, B.; Bruno, C.; Cauzillo, G.; Cavone, D.; Convertini, L.; De Mei, B.; Marconi, A.; Montagano, G.; Musti, M.; et al. Mesotelioma pleurico ed esposizione ambientale a fibre minerali: Il caso di un'area rurale in Basilicata. *Ann. Ist. Super. Sanita.* **2004**, *40*, 251–265. [PubMed]

30. Punturo, R.; Ricchiuti, C.; Mengel, K.; Apollaro, C.; De Rosa, R.; Bloise, A. Serpentinite-derived soils in southern Italy: Potential for hazardous exposure. *J. Mediterr. Earth Sci.* **2018**, *10*, 51–61.

31. Monaco, C.; Tortorici, L. Tettonica estensionale quaternaria nell'Arco Calabro e in Sicilia orientale. *Studi Geologici Camerti* **1995**, *2*, 351–362.

32. Monaco, C.; Tortorici, L.; Paltrinieri, W. Structural evolution of the Lucanian Apennines, southern Italy. *J. Struct. Geol.* **1998**, *20*, 617–638. [CrossRef]

33. Vezzani, L. La Formazione del Frido (Neocomiano- Aptiano) tra il Pollino e il Sinni. *Geol. Rom.* **1969**, *8*, 129–176.

34. Amodio Morelli, L.; Bonardi, G.; Colonna, V.; Dietrich, D.; Giunta, G.; Ippolito, F.; Liguori, V.; Lorenzoni, S.; Paglioncino, A.; Perrone, V. L' arco Calabro Peloritano nell' orogene Appenninico-Maghrebide. *Mem. Soc. Geol. It.* **1976**, *17*, 1–60.

35. Vezzani, L. La sezione tortoniana di Perosa sul fiume Sinni presso Episcopia (Potenza). *Geol. Rom.* **1966**, *5*, 263–290.

36. Sansone, M.T.C.; Rizzo, G.; Mongelli, G. Petrochemical characterization of mafic rocks from the Ligurian ophiolites, Southern Apennines. *Int. Geol. Rev.* **2011**, *53*, 130–156. [CrossRef]

37. Rizzo, G.; Cristi Sansone, M.T.; Perri, F.; Laurita, S. Mineralogy and petrology of the metasedimentary rocks from the frido unit (southern apennines, Italy). *Period. Mineral.* **2016**, *85*, 153–168.

38. Sansone, M.T.C.; Prosser, G.; Rizzo, G.; Tartarotti, P. Spinel-peridotites of the frido unit ophiolites (southern apennine-italy): Evidence for oceanic evolution. *Period. Mineral.* **2012**, *81*, 35–59.

39. Whitney, D.L.; Evans, B.W. Abbreviations for names of rock-forming minerals. *Am. Mineral.* **2010**, *95*, 185–187. [CrossRef]

40. Leake, B.E.; Woolley, A.R.; Arps, C.E.S.; Birch, W.D.; Gilbert, M.C.; Grice, J.D.; Hawthorne, F.C.; Kato, A.; Kisch, H.J.; Krivovichev, V.G.; et al. Nomenclature of amphiboles: Report of the subcommittee on amphiboles of the international mineralogical association, commission on new minerals and mineral names. *Can. Mineral.* **1997**, *35*, 219–246.

41. Bloise, A.; Barrese, E.; Apollaro, C.; Miriello, D. Flux growth and characterization of Ti and Ni doped forsterite single crystals. *Cryst. Res. Technol.* **2009**, *44*, 463–468. [CrossRef]

42. Bloise, A.; Catalano, M.; Gualtieri, A.F. Effect of Grinding on Chrysotile, Amosite and Crocidolite and Implications for Thermal Treatment. *Minerals* **2018**, *8*, 135. [CrossRef]

43. Bloise, A.; Kusiorowski, R.; Gualtieri, A.F. The Effect of Grinding on Tremolite Asbestos and Anthophyllite Asbestos. *Minerals* **2018**, *8*, 274. [CrossRef]

44. Ballirano, P.; Bloise, A.; Gualtieri, A.F.; Lezzerini, M.; Pacella, A.; Perchiazzi, N.; Dogan, M.; Dogan, A.U. The Crystal Structure of Mineral Fibers. In *Mineral Fibers: Crystal Chemistry, Chemical-Physical Properties, Biological Interaction and Toxicity*; Gualtieri, A.F., Ed.; European Mineralogical Union: London, UK, 2017; Volume 18, pp. 17–53.

45. Bloise, A.; Kusiorowski, R.; Lassinantti Gualtieri, M.; Gualtieri, A.F. Thermal behaviour of mineral fibers. In *Mineral Fibers: Crystal Chemistry, Chemical-Physical Properties, Biological Interaction and Toxicity*; Gualtier, A.F., Ed.; European Mineralogical Union: London, UK, 2017; Volume 18, pp. 215–252.

46. Loomis, D.; Dement, J.; Richardson, D.; Wolf, S. Asbestos fibre dimensions and lung cancer mortality among workers exposed to chrysotile. *Occup. Environ. Med.* **2010**, *67*, 580–584. [CrossRef] [PubMed]

47. Suzuki, Y.; Yuen, S.R.; Ashley, R. Short, thin asbestos fibersc ontribute to the development of human malignant mesothelioma: Pathological evidence. *Int. J. Hyg. Environ. Health* **2005**, *208*, 201–210. [CrossRef] [PubMed]

48. Bernstein, D.; Castranova, V.; Donaldson, K.; Fubini, B.; Hadley, J.; Hesterberg, T.; Kane, A.; Lai, D.; McConnell, E.E.; Muhle, H.; et al. Testing of fibrous particles: Short-term assays and strategies. *Inhal. Toxicol.* **2005**, *17*, 497–537. [CrossRef] [PubMed]

49. Stanton, M.F.; Layard, M.; Tegeris, A.; Miller, E.; May, M.; Morgan, E.; Smith, A. Relation of particle dimension to carcinogenicity in amphibole asbestoses and other fibrous mineral. *J. Natl. Cancer. Inst.* **1981**, *67*, 965–975. [PubMed]

50. Dichicco, M.C.; Laurita, S.; Sinisi, R.; Battiloro, R.; Rizzo, G. Environmental and Health: The Importance of Tremolite Occurence in the Pollino Geopark (Southern Italy). *Geosciences* **2018**, *8*, 98. [CrossRef]

51. Bellomo, D.; Gargano, C.; Guercio, A.; Punturo, R.; Rimoldi, B. Workers' risks in asbestos contaminated natural sites. *J. Mediterr. Earth Sci.* **2018**, *10*, 97–106.

fibers

MDPI

Article

Mineralogical and Microstructural Features of Namibia Marbles: Insights about Tremolite Related to Natural Asbestos Occurrences

Rosalda Punturo [1],*, Claudia Ricchiuti [1], Marzia Rizzo [2] and Elena Marrocchino [2]

[1] Department of Biological, Geological and Environmental Sciences, University of Catania, Corso Italia 55, 95129 Catania, Italy; claudia.ricchiuti@unict.it
[2] Department of Physics and Earth Science—University of Ferrara, Via Saragat 1, 44122 Ferrara, Italy; marzia.rizzo@unife.it (M.R.); mrrlne@unife.it (E.M.)
*   Correspondence: punturo@unict.it; Tel.: +39-095-719-5757

Received: 27 February 2019; Accepted: 28 March 2019; Published: 7 April 2019

Abstract: The Mg-rich marbles of Precambrian rocks of Namibia are widely exploited and marketed abroad for ornamental purposes. Karibib marbles, named after the locality where the most important quarries are located, are commercially known as "White Rhino Marble". They formed under greenschist facies metamorphic conditions and may be characterized by the presence of veins of tremolite. Although the quarries, whose exploited marbles contain tremolite, do not seem to be abundant, we decided to carry out a detailed mineralogical and petrographic study on Karibib marbles in order to point out the occurrence of tremolite, whose shape may vary from prismatic to acicular, even sometimes resembling the asbestiform habitus and its geometry within the rock. With this aim, we carried out optical microscopy, X-ray diffractometry, X-ray scanning electron microscopy, and micro-Raman investigations, and also imaged the 3D fabric with micro computed X-ray tomography. The study of white marbles from Namibia and their mineral phases has an important impact, since tremolite might split into thin fibers and, therefore, being potentially harmful, the presence of tremolite requires an analysis of the risks of exposure to asbestos.

Keywords: rhino white marble of Namibia; tremolite; fibrous habitus; health hazard

1. Introduction

In the last century, Namibia has been one of the favorable mining contexts for the exploration and evaluation of geo-resources. From 1990 to 2000, in Namibia, the production of marble and granite was about 20,000 tons per year. Since 2004, thanks to modern methods and processing machinery, there has been a continuous increase in production, and the production has exceeded the threshold of 50,000 tons/year [1]. The geo-mining industry of Namibia includes several ornamental stones: marbles (calcite and/or dolomite-bearing metacarbonate rocks); magmatic rocks such as granites, granodiorites, and gabbros; serpentinites; and onyxes and alabasters. The firms linked to the ornamental stone that are gathered around the Karibib have become a benchmark for the high quantity and quality of marble.

Since 1900, railway construction has led to great development in mining and, in the past 10 years, the Karibib has become one of the most productive international marble districts that includes extraction, processing, and marketing activities of marble and granite rocks.

In the Karibib district, the most important marble and granite reserves are located in the Karibib quarry area of the northwest sector of town, where the White Rhino and Karibib marble varieties are exploited; in the Nonidas quarry area, which consists of small extractive sites that sit between the northern part of the town of Nonidas and the eastern area of Swakopmund; and the Arandis quarry area, where the extraction activities mainly concern the domes of intrusive magmatic rocks (pink granite).

At present, there is growing interest due to the ornamental exploitation of these Neoproterozoic carbonate rocks, and many quarries are contributing to the socio-economic development of Namibia and other regions—indeed, extensive outcrops of carbonate rocks are part of Namibia's geological resources and are therefore recalling the interest of mining companies (see also the website of the Namibian Ministry of Mines and Energy [2]).

In this work, mineralogical and petrographic characteristics of the main commercial marble in the Karibib area, known as "Rhino White Marble", are described. It is a dolomite-bearing marble from the Neoproterozoic, which belongs to the Swakop group (Damara sequence). It is exploited and marketed in many European countries, and it is appreciated because of its pearly white appearance, sometimes cut by creamy yellow veins. However, some concerns related to the commercial use of Rhino White Marble are due to the occurrence of tremolite-rich veins, as revealed by preliminary petrographic investigation [3]. Indeed, tremolite, $Ca_2Mg_5Si_8O_{22}(OH)_2$, belongs to the calcic amphibole group of minerals and, when occurring with fibrous habitus, it is considered a dangerous naturally-occurring asbestos—a term applied to six specific silicate minerals that also comprises tremolite—the critical dimension is: length > 5 μm, diameter < 3 μm, length:diameter > 3:1 [4–6].

This mineral usually occurs with elongated and/or bladed prismatic habitus, but it may also be acicular or even fibrous-shaped. According to the literature, tremolite toxicology, as for all asbestos minerals, has been associated with size, durability, and chemical composition (e.g., [7–15]). According to [16], "In mineralogy, acicular is the term applied to straight, free-standing (i.e., individual) and highly elongated crystals; these ones can be bordered and delimited by crystal faces. As far as the acicular crystals, they are characterized by aspect ratio comparable to those ones of fibrous crystals, even though their diameter may extend up to 7 mm". A fiber is defined as an elongate particle that is longer than 5.0 μm, with a minimum aspect ratio (length of the particle divided by its width) of 3:1 [6]. Indeed, when used as building stone, the studied marbles are washed with aggressive detergents and also exposed to accelerated weathering, so the mineral fibers contained within could break and may be spread out in the environment and make them dangerous for the environment and human health [17–26].

Although the quarries of the Karibib area that sit on tremolite-bearing marbles do not seem to be abundant, we considered it necessary to carry out a detailed mineralogical and microstructural investigation in order to characterize the white marbles of Namibia and to detect the eventual occurrence of asbestos tremolite. For the above reasons, the present study has several implications, since the presence of tremolite with asbestiform habitus might be linked to health problems and asbestosis. Therefore, it is a useful tool for initiating an analysis of the risks to occupational and non-occupational activities concerning the use of the tremolite-bearing marble, providing useful suggestions for safe marble exploitation.

2. Materials and Methods

2.1. Geological Setting and Samples

The Namibia marbles belong to the Neoproterozoic carbonate succession, dating 665 ± 34 million years, which constitute the Pan-African Damara Belt. The latter was generated during the orogenic events that produced the Gondwana supercontinent. The sedimentary successions of the Damara Belt, siliciclastic and carbonate in composition, were deposited in an environment of passive continental margin (i.e., Neoproterozoic rift basins) related to the Rodinia break-up on a global scale. In some sectors, the thickness of deposits exceeds 1000 m [27–30]. According to the literature [31–34], the Damara Belt is considered as an asymmetric double-vergent orogen, which separates the Angola-Congo and Kalahari cratons (Figure 1), formed during the Neoproterozoic to early Paleozoic tectonic events related to the closure of the Damara Ocean. In Namibia, the Damara

Orogen (Figure 1) is constituted by three orogenic belts: The intracontinental Damara belt and the coastal belt, the Kaoko belt, and the Gariep belts [35–38].

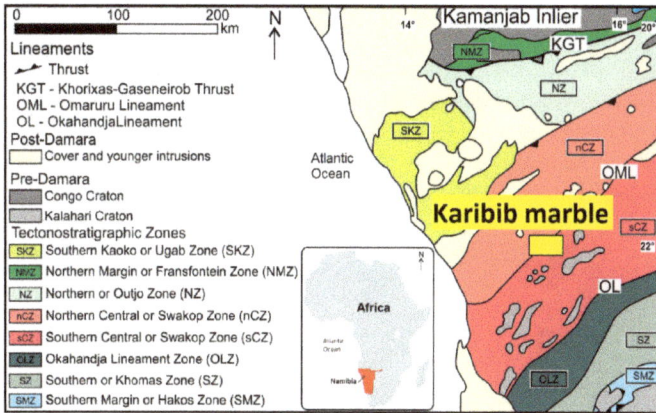

Figure 1. Simplified tectonic map of the Damara Belt (Namibia (Africa). showing the distribution of the main tectono-stratigraphic zones according to [28]. Modified after [35].

In the Central Zone of the Damara Belt, the successions were deformed and metamorphosed to greenschist facies conditions reaching, in some sectors, metamorphic conditions of up to ca. 590 °C and 0.5 GPa [39]. Moreover, a detailed structural mapping [40–45], highlighted as the most striking structural feature of the Central zone, is the northeast trending domes elongated at kilometer-scale [37,40,42,45,46], where the most important quarries are located (Figure 2).

Figure 2. Schematic map showing the northeast trending dome structures covered by the geological map, by [40]. The gneisses and/or the Pan-African granitoids constitute the cores of the dome structures, whereas the surrounding supracrustals of the Damara sequence are draped around the domes. The solid black arrows indicate the tectonic transport direction for domes [41–44] and the Karibib district in the northeast. The yellow rectangle indicates the area where the marble quarries are located.

Deformation and metamorphism have not completely obliterated structures and textures inherited from sedimentary environments, so the planar surfaces due to tectonic deformation overprint the contacts between lithofacies [35,47–49].

Within the carbonate protolith, which represents a pelagic environment with main carbonate sedimentation and a lower contribution of siliceous organisms in relation to the oscillations of Carbonate Compensation Depth, the derived magnesium-rich marbles—which underwent greenschist facies metamorphism—can be characterized by the occurrence of tremolite as one of the main constituting minerals. Conversely, the portions closest to the continental margins do not have tremolite because the Al and Fe terrigenous sediments, metamorphosed under greenschist facies conditions, give chlorite in the metamorphic assemblage.

At the scale of the quarry, White Rhino Marbles of the Karibib area look pearly white and are extracted as dimension stones (Figure 3a)—at the mesoscopic scale, marbles show a saccaroid fabric and are cut by yellowish veins (Figure 3b,c).

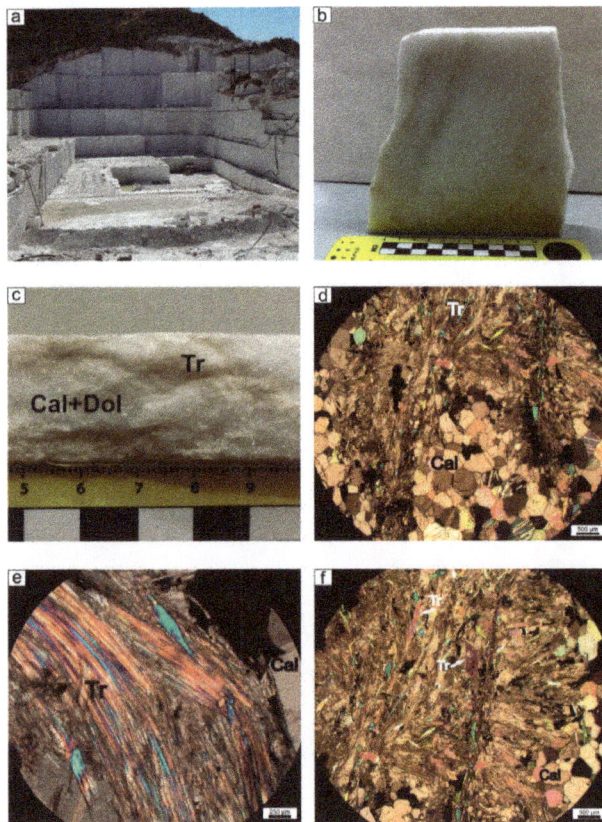

Figure 3. The main features of Karibib white marbles. (**a**) Front of a quarry where marble is exploited as dimensional stone; (**b**) Appearance of the marble at the mesoscopic scale, note the yellow-greenish veins across the pearly portion; (**c**) Particular of a cross section of a brick; (**d–f**): Photomicrographs of thin sections of marbles; (**d**) Coexistence of granoblastic portions constituted by calcite with nematoblastic portions constituted by tremolite; (**e**) Blow-up of acicular tremolite-rich level; (**f**) Nematoblastic level showing various habitus types of tremolite, from prismatic to acicular. Mineral symbols after [50].

2.2. *Methodologies*

In order to describe the microstructural features of the investigated marbles, we selected some specimens for optical microscopy (OM), scanning electron microscopy (SEM/EDS), X-ray diffractometry (XRD), micro-Raman spectrometry, and synchrotron radiation X-ray microtomography

(SR X-ray μCT) investigations. A polarizing microscope Zeiss Axiolab and a Tescan-Vega\\LMU scanning electron microscope (Tescan-Vega, Brno – Kohoutovice, Czech Republic) equipped with an Edax Neptune XM4 60 energy-dispersive X-ray spectrometer (EDS), Edax, Mahwah, NJ, 07430 USA) operating at 20 kV accelerating voltage and 20 nA beam current conditions, were employed to obtain microstructural features, morphoscopic images, and elemental microanalyses. Investigation was carried out on polished thin sections as well as on small chips of marble specimens.

Some specimens were also examined through the X-ray diffractometry (XRD) technique to establish the mineralogical composition. The XRD analysis was performed on rock powder using a Philips PW1860/00 diffractometer (Philips Panalytical Canton, MA, USA), with graphite-filtered Cu Kα radiation (1.54 Å), allowing determination of the mineralogical phases within the constituents. Diffraction patterns were collected in the 2θ angular range 5–50°, with 5 s/step (0.02° 2θ). Moreover, XRD data were quantified by the RIR (Reference Intensity Ratio) method of powder X-ray diffraction data in order to establish the quantities of the constituting minerals according to [51].

A LabRam HR800 micro-Raman instrument from Horiba Jobin Yvon (Horiba, Kyoto, Japan), equipped with an air-cooled CCD detector (1024 × 256 pixels) at −70 °C, an Olympus BXFM microscope, a 600 groove/mm grating, and a 50× objective, was used to collect the Raman scattering signals. The excitation source was a He–Ne laser (632.8 nm line) whose maximum power was 20 mW. The spectrometer was calibrated with silicon at 520 cm^{-1} and the exposure time was varied from 50 to 100 s. Data obtained were compared with the RRUFF^{TM} project database [52]. Moreover, one selected sample considered to be representative of the microstructural features of marbles was imaged by synchrotron radiation X-ray microtomography (SR X-ray μCT) at the SYRMEP (SYnchrotron Radiation for MEdical Physics) beamline of the Elettra synchrotron (Elettra - Sincrotrone Trieste S.C.p.A, Trieste, Italy) in white-beam configuration mode at high spatial resolution. To this aim, we cut a parallelepiped with a size of about 4 mm. The X-ray spectrum was filtered for low energies with 1 mm of Si + 1 mm of Al, and the sample-to-detector distance was set to 200 mm. For each measurement, 1800 projections were acquired over a total scan angle of 180° with an exposure time/projection of 2 s. The detector consisted of a 16-bit air-cooled sCMOS camera (Hamamatsu C11440 22C, Hamamatsu City, Japan) with a 2048°—2048° pixels chip. The effective pixel size of the detector was set at 1.952 μm^2, yielding a maximum field of view of ca. 3.22 mm^2. Since the lateral size of the samples was larger than the detector field of view, the X-ray tomographic microscans were acquired in local or region-of-interest mode [53]. A single distance phase retrieval-preprocessing algorithm [54] was applied to the white beam projections in order to improve the reliability of the quantitative morphological analysis and enhance the image contrast.

The obtained 3D volumes were then imported in VGStudio Max 2.2 (Volume Graphics, Charlotte, NC, USA) for the 3D rendering and segmentation by manual thresholding.

3. Results

At the scale of the microscope, the White Rhino Marbles of the Karibib area had relatively fine grain size with a very heterogeneous distribution of white and yellowish levels, as was revealed by previous mesoscopic observation (Figure 3b,c). Indeed, two main domains, whose thickness ranged from 2 mm up to 1–2 cm, were distinguished on the basis of evident microstructures and constituting minerals: granoblastic levels, given by calcite +/− dolomite (Figure 3d), which are the most abundant portions of the rocks, as also highlighted by mesoscopic observations. Conversely, the nematoblastic levels that occurred to a minor extent, were characterized by tremolite, occurring with various habitus types—indeed, there were levels in which tremolite crystals were made of well-developed prismatic to acicular minerals and minor levels in which this mineral phase tended to constitute fiber belts (Figure 3e,f).

The granoblastic portions, prevalently constituted by calcite and dolomite, showed straight grain boundaries. They were also sutured, even embayed, resulting in an interlocked texture (Figure 3d)—the greenish-yellowish nematoblastic levels, showing marked microstructural anisotropy, were given by tremolite, occurring as either acicular crystals and/or highly elongated fiber aggregates in belts with radial disposition (Figure 3d,f), together with hetero-granoblastic calcite and dolomite grains elongated parallel to foliation.

XRD and micro-Raman analyses (Figures 4 and 5) showed the coexistence of calcite, tremolite, and dolomite and the absence of other mineral phases also in the finest-grained nematoblastic portions of the yellowish bands, without secondary or accessory minerals occurring. Quantitative phase analyses with the RIR method showed that, on average, the abundances of constituent minerals determined on powders obtained from representative bricks were calcite 70%, tremolite 26%, and dolomite 4%.

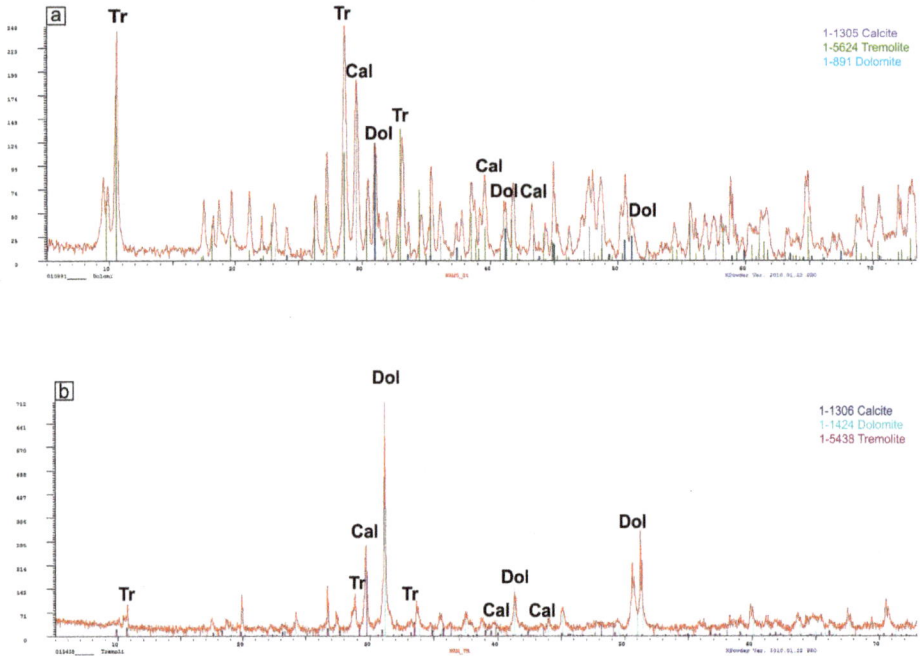

Figure 4. X-ray diffractograms on representative portions of the Karibib white marbles, showing calcite, tremolite, and dolomite as constituting mineral phases. The peak color of each mineral is indicated in the legend. (**a**) main vein; (**b**) massive part.

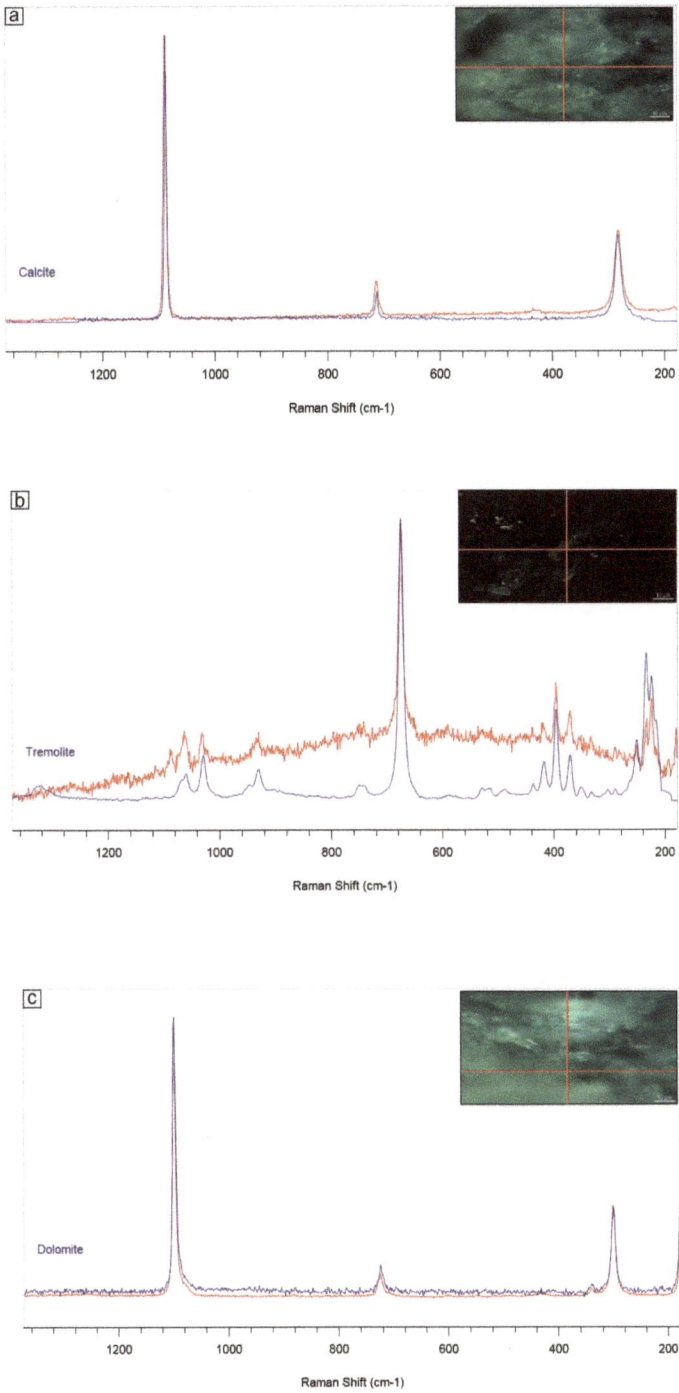

Figure 5. Micro-Raman spectra of selected areas of the Karibib white marbles, showing (**a**) calcite; (**b**) tremolite; and (**c**) dolomite as constituting mineral phases (red spectra). The blue reference spectra are after [52]. Pictures of boxes indicate the investigated points.

The rock exhibited a friable appearance, especially at the contact areas between amphibole and carbonate minerals, with tremolite occurring either with prismatic elongated habitus or elongated fibers, closely bound to carbonate minerals, as can be seen in the SEM images (Figure 6a–c). The SEM/EDS elemental microanalysis suggested that tremolite individuals were pure Mg-member $Ca_2Mg_5(OH)_2Si_8O_{22}$ without any iron detected [55]. Moreover, the SEM images did show that, as a consequence of disaggregation, tremolite might also split into fibers and cleavage fragments, whose shape parameters may resemble asbestiform habitus (Figure 6d–f).

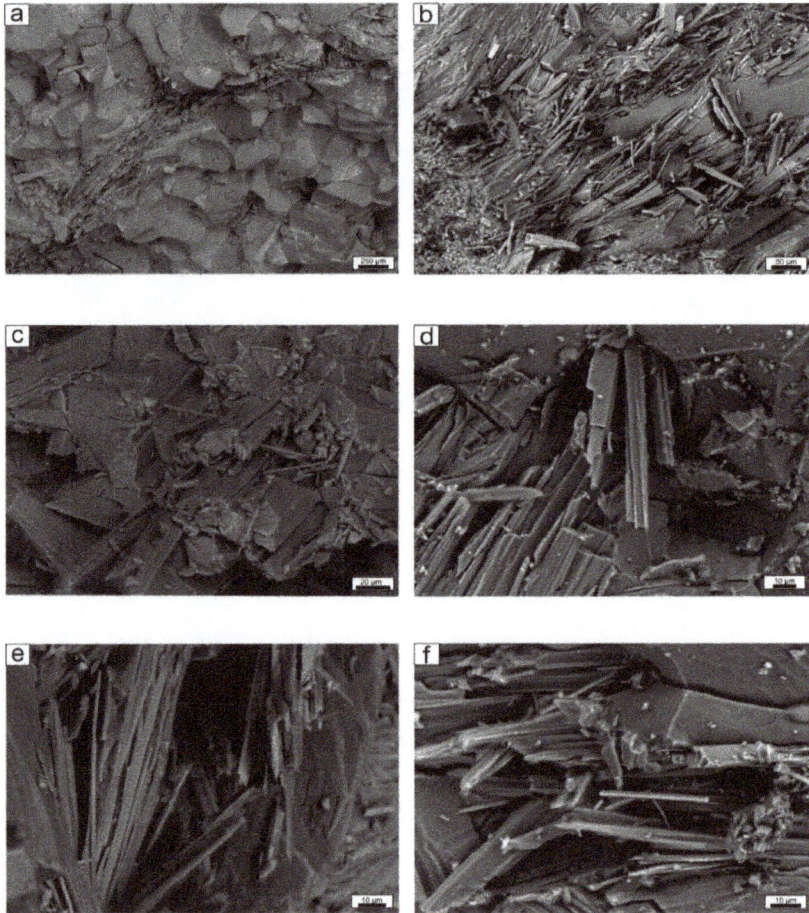

Figure 6. Scanning electron photomicrographs. (**a**) Calcite-rich granoblastic levels cut by tremolite veins; (**b**) Elongated tremolite crystals, sometimes showing radial disposition; (**c**) Tremolite splitting into fibers; (**d**) Tremolite cleavage fragments prone to split; (**e,f**) tremolite fibers whose shape parameters may resemble asbestiform habitus. Mineral symbols after [50].

Finally, on one selected small brick measuring about 30 mm × 4 mm, we carried out synchrotron radiation X-ray microtomography (SR X-ray μCT). X-ray microtomography is a non-destructive technique that improves the observation of the arrangement of fibers in the three-dimensional space, thus avoiding any morphological variations of the sample as a result of comminution. Indeed, this technique allowed us to image the three-dimensional enveloping and intergrowth of nematoblastic and granoblastic levels as well as the geometry and reciprocal arrangement of constituting minerals into the marble, with special regard to the spatial relationship between calcite and tremolite, the latter sometimes showing radial disposition, as can be clearly seen in Figure 7.

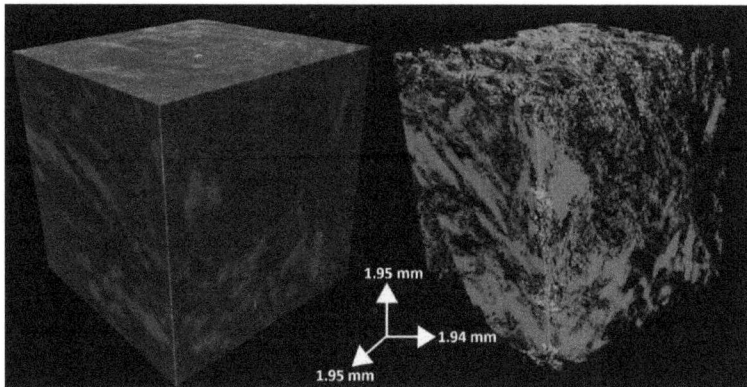

Figure 7. 3D rendering of a selected part of one specimen analyzed by means of synchrotron radiation X-ray micro-tomography: in the left picture, light colors correspond to high-density phases, while dark colors correspond to low-density phases; in the right picture, a green color is associated with the highest-density phase (i.e., tremolite). Note the 3D interlock between tremolite- and carbonate-rich portions.

4. Discussion and Conclusions

The multi-analytical investigation carried out on White Rhino Marbles exploited in the Karibib area (Namibia), which consisted of a detailed petrographic, microstructural, and mineralogical characterization of their fabric and microstructural features, permitted us to highlight the occurrence and to depict the geometry of amphibole minerals in the yellow veins that cut the rock. From the petrological point of view, the Neoproterozoic White Rhino Marbles are characterized by a mineralogical assemblage that proves the absence of terrigenous contributions in their protolith, as they do not contain any aluminum or iron, even in nematoblastic levels in which silicate mineral phases (i.e., amphibole) are found.

During the metamorphic event, the high-silica (e.g., diatomaceous) levels reacted with the Mg-rich carbonates, giving rise to amphibole tremolite $Ca_2Mg_5Si_8O_{22}(OH)_2$. Therefore, the paragenesis of the White Rhino Marbles is given by calcite + tremolite $\pm$ dolomite. Calcite and minor dolomite grains are the constituent of the granoblastic levels, which are certainly the most abundant portions of the marble rocks exploited. Conversely, tremolite is the principal constituent of the nematoblastic levels, where it is mainly found with acicular (i.e., needle-like) habitus, which means it is characterized by sectional dimensions that are small relative to its length. Moreover, no secondary minerals formed on primary minerals have been detected or observed, proving that no weathering process has been affecting the studied marbles.

Nevertheless, the detailed microstructural and morphological analyses carried out on marbles highlighted that, despite non-asbestos tremolite exhibiting acicular habitus, it is the most common mineral phase that was found. Asbestos tremolite fibers were also detected within veins. Tremolite-rich veins were easy to distinguish at the mesoscopic and at the optical microscopic scale, where they defined the microstructural anisotropy of marbles. Scanning the electron microscopy highlighted that tremolite fibers, resembling the asbestiform habitus, occurred as fibrous aggregates with radial arrangement, prone to split into thinner fibers and ultimately into fibrils, often formed after cleavage fragments. Despite its occurring habitus, tremolite appeared as straight and stiff crystals (i.e., needles and fibers). Moreover, 3D imaging showed the tight interlock between the nematoblastic microdomains (i.e., tremolite-rich) and the granoblastic portions (i.e., carbonate-rich) and their contact geometry.

The asbestos hazard related to the occurrence of fibrous tremolite veins that cross-cut the studied marbles arises when either natural weathering processes (e.g., erosion and mobilization) or human activities (e.g., exploitation of dimension blocks and subsequent use as building stones) separate tremolite fibers and break them down, making them dispersed into the environment as airborne and easily breathable.

For instance, during the steps of marble quarrying, non-asbestos tremolite can break along preferred cleavage planes and be released in the air. For this reason, it is ultimately possible for workers to be exposed to asbestos during these activities. Therefore, before any exploitation and subsequent process of marble containing non-asbestos minerals, which may otherwise develop into minerals with asbestiform habitus, it is necessary that mining companies adopt monitoring surveys, in situ tests, as well as safety measures and prevention practices for each recognized hazardous situation. Among them it is worth noting the avoidance of asbestos veins during exploitation, mainteinance of devices, use of protective personal equipment, planning sanitary surveillance, and envisaging dust abatement and remediation systems [21,56–58]. As far as the non-occupational point of view states, it is important to assess the extent of exposure to those airborne particles, whose morphology may resemble asbestos, to populations who live close to the quarry as well as to family members of workers. Finally, we suggest that weathering and ageing tests should be carried out on vein-rich marble, in order to detect any deterioration forms that may cause the release of fibers, and to plan eventual remediation practices.

Author Contributions: Each author made substantial individual contribution to the work as follows: Conceptualization, R.P.; methodology and analysis, R.P., C.R., and M.R.; writing, R.P., C.R., M.R., and E.M.; validation, R.P. and E.M.; editing C.R. and E.M.; funding acquisition R.P.

Funding: Part of this research was carried out under the financial support of Piano Triennale della Ricerca (2017–2020 and, later, Università di Catania, Dipartimento di Scienze Biologiche, Geologiche e Ambientali), "L'amianto naturale nelle rocce e nei suoli: implicazioni ambientali e relazioni con le attività umane". Scientific responsibility: Rosalda Punturo.

Acknowledgments: The authors are grateful to Elettra Synchrotrone and to Lucia Mancini (Trieste, Italy) for the SYRMEP facilities. Technical support by Gabriele Lanzafame is also acknowledged. The authors are grateful to two anonymous reviewers for constructive comments that improved the manuscript. Suggestions by Andrea Bloise are appreciated. Editorial handling by Billy Bai is acknowledged.

Conflicts of Interest: The authors declare no conflict of interest.

References

1. Ministry of Mines and Energy. Available online: http://www.mme.gov.na (accessed on 20 January 2019).
2. Ministry of Mines and Energy. Available online: http://www.mme.gov.na/directorates/mine/ (accessed on 20 January 2019).
3. Vaccaro, C.; Punturo, R.; Volpe, L.; Marrocchino, E.; Rizzo, M.; Ricchiuti, C.; Mannino, M. *Tremolite—Bearing Marbles from Namibia Exploited as Dimension Stone: Preliminary Mineralogical and Petrographic Characterization*; Abstract Book; Congresso SGI-SIMP: Catania, Italy, 2018.
4. World Health Organization (WHO). *Asbestos and Other Natural Mineral Fibres*; Environmental Health Criteria, 53; World Health Organization: Geneva, Switzerland, 1986; p. 194.
5. National Institute for Occupational Safety and Health (NIOSH). *Asbestos and Other Elongated Mineral Particles: State of the Science and Roadmap for Research*; Revised Draft; NIOSH Current Intelligence Bulletin: Washington, DC, USA, 2008.
6. International Agency for Research on Cancer (IARC). *Asbestos (Chrysotile, Amosite, Crocidolite, Tremolite, Actinolite, and Anthophyllite) IARC Monographs*; Arsenic, Metals, Fibres and Dusts; International Agency for Research on Cancer: Lyon, France, 2009; pp. 147–167.
7. Mossman, B.T.; Lippmann, M.; Hesterberg, T.W.; Kelsey, K.T.; Barchowsky, A.; Bonner, J.C. Pulmonary endpoints (lung carcinomas and asbestosis) following inhalation exposure to asbestos. *J. Toxicol. Environ. Health* **2011**, *14*, 76–121. [CrossRef] [PubMed]

8. Pugnaloni, A.; Giantomassi, F.; Lucarini, G.; Capella, S.; Bloise, A.; Di Primio, R.; Belluso, E. Cytotoxicity induced by exposure to natural and synthetic tremolite asbestos: An in vitro pilot study. *Acta Histochem.* **2013**, *115*, 100–112. [CrossRef] [PubMed]

9. Oberti, R.; Hawthorne, F.C.; Cannillo, E.; Camara, F. Long-range order in amphiboles. In *Amphiboles: Crystal Chemistry, Occurrence, and Health Issues Hawthorne*; Oberti, F.C., della Ventura, R., Mottana, G.A., Eds.; Mineralogical Society of America: Chantilly, VA, USA, 2007; pp. 125–172.

10. Ballirano, P.; Bloise, A.; Gualtieri, A.F.; Lezzerini, M.; Pacella, A.; Perchiazzi, N.; Dogan, M.; Dogan, A.U. The Crystal Structure of Mineral Fibres. In *Mineral Fibres: Crystal Chemistry, Chemical-Physical Properties, Biological Interaction and Toxicity*; Gualtieri, A.F., Ed.; European Mineralogical Union: London, UK, 2017; Volume 18, pp. 17–53.

11. Gaggero, L.; Sanguineti, E.; Yus González, A.; Militello, G.M.; Scuderi, A.; Parisi, G. Airborne asbestos fibres monitoring in tunnel excavation. *J. Environ. Manag.* **2017**, *196*, 583–593. [CrossRef] [PubMed]

12. Apollaro, C.; Fuoco, I.; Vespasiano, G.; De Rosa, R.; Cofone, F.; Miriello, D.; Bloise, A. Geochemical and mineralogical characterization of tremolite asbestos contained in the Gimigliano-Mount Reventino Unit (Calabria, south Italy). *JMES* **2018**, *10*, 5–15. [CrossRef]

13. Punturo, R.; Bloise, A.; Critelli, T.; Catalano, M.; Fazio, E.; Apollaro, C. Environmental implications related to natural asbestos occurrences in the ophiolites of the Gimigliano-Mount Reventino Unit (Calabria, southern Italy). *Int. J. Environ. Res.* **2015**, *9*, 405–418.

14. Constantopoulos, S.H. Environmental mesothelioma associated with tremolite asbestos: Lessons from the experiences of Turkey, Greece, Corsica, New Caledonia and Cyprus. *Regul. Toxicol. Pharmacol.* **2008**, *52*, 110–115. [CrossRef] [PubMed]

15. Bloise, A.; Barca, D.; Gualtieri, A.F.; Pollastri, S.; Belluso, E. Trace elements in hazardous mineral fibres. *Environ. Pollut.* **2016**, *216*, 314–323. [CrossRef]

16. Strohmeier, B.R.; Huntington, J.C.; Bunker, K.L.; Sanchez, M.S.; Allison, K.; Lee, R.J. What is asbestos and why is it important? Challenges of defining and characterizing asbestos. *Int. Geol. Rev.* **2010**, *52*, 801–872. [CrossRef]

17. Petriglieri, J.R.; Laporte-Magoni, C.; Gunkel-Grillon, P.; Tribaudino, M.; Bersani, D.; Sala, O.; Salvioli-Mariani, E. Mineral fibres and environmental monitoring: A comparison of different analytical strategies in new caledonia. *Geosci. Front.* **2019**. [CrossRef]

18. Laporte-Magoni, C.; Petriglieri, J.R.; Gunkel-Grillon, P.; Salvioli-Mariani, E.; Selmaoui-Folcher, N.; Le Mestre, M. Weathering Influence on fiber release of asbestos type minerals under subtropical climate. In Proceedings of the XXII Meeting of the International Mineralogical Association, Melbourne, Australia, 13–17 August 2018.

19. Bloise, A.; Kusiorowski, R.; Lassinantti Gualtieri, M.; Gualtieri, A.F. Thermal behaviour of mineral fibres. In *Mineral Fibres: Crystal Chemistry, Chemical-Physical Properties, Biological Interaction and Toxicity*; Gualtier, A.F., Ed.; European Mineralogical Union: London, UK, 2017; Volume 18, pp. 215–252.

20. Bloise, A.; Punturo, R.; Catalano, M.; Miriello, D.; Cirrincione, R. Naturally occurring asbestos (NOA) in rock and soil and relation with human activities: The monitoring example of selected sites in Calabria (southern Italy). *Ital. J. Geosci.* **2016**, *135*, 268–279. [CrossRef]

21. Bellomo, D.; Gargano, C.; Guercio, A.; Punturo, R.; Rimoldi, B. Workers' risks in asbestos contaminated natural sites. *JMES* **2018**, *10*, 97–106.

22. Pacella, A.; Andreozzi, G.B.; Ballirano, P.; Gianfagna, A. Crystal chemical and structural characterization of fibrous tremolite from Ala di Stura (Lanzo Valley, Italy). *Period. Mineral.* **2008**, *77*, 51–62.

23. Belardi, G.; Vignaroli, G.; Trapasso, F.; Pacella, A.; Passeri, D. Detecting asbestos fibres and cleavage fragments produced after mechanical tests on ophiolite rocks: Clues for the asbestos hazard evaluation. *JMES* **2018**, *10*, 63–78. [CrossRef]

24. Punturo, R.; Cirrincione, R.; Pappalardo, G.; Mineo, S.; Fazio, E.; Bloise, A. Preliminary laboratory characterization of serpentinite rocks from Calabria (southern Italy) employed as stone material. *J. Mediterr. Earth Sci.* **2018**, *10*, 79–87. [CrossRef]

25. Punturo, R.; Ricchiuti, C.; Mengel, K.; Apollaro, C.; De Rosa, R.; Bloise, A. Serpentinite-derived soils in southern Italy: Potential for hazardous exposure. *J. Mediterr. Earth Sci.* **2018**, *10*, 51–61. [CrossRef]

26. Censi, P.; Zuddas, P.; Randazzo, L.A.; Tamburo, E.; Speziale, S.; Cuttitta, A.; Punturo, R.; Santagata, R. Source and nature of inhaled atmospheric dust from trace element analyses of human bronchial fluids. *Environ. Sci. Technol.* **2011**, *45*, 6262–6267. [CrossRef]

27. South African Committee for Stratigraphy, SACS. Stratigraphy of South Africa, Part I. Lithostratigraphy of the Republic of South Africa, South West Africa/Namibia and the Republics of Bophuthatswana, Transkei and Venda. *Geol. Surv. S. Afr.* **1980**, *8*, 690.

28. Kröner, A. Late Precambrian diamictites of South Africa and Namibia. In *Earth's Pre-Pleistocene Glacial Record*; Hambrey, M.J., Harland, W.B., Eds.; Cambridge University Press: Cambridge, UK, 1981; pp. 167–177.

29. Paciullo, F.V.P.; Ribeiro, A.; Trouw, R.A.J.; Passchier, C.W. Facies and facies association of the siliciclastic Brak River and carbonate Gemsbok formations in the Lower Ugab River valley, Namibia, W. *Africa. J. Afr. Earth Sci.* **2007**, *47*, 121–134. [CrossRef]

30. Miller, R.M. Neoproterozoic and early Palaeozoic rocks of the Damara Orogen. In *The Geology of Namibia*, 2nd ed.; Miller, R.M., Ed.; Geological Survey of Namibia: Windhoek, Namibia, 2008; pp. 1311–1341.

31. Hoffmann, K.H. *Sedimentary Depositional History of the Damara Belt Related to Continental Breakup, Passive to Active Margin Transition and Foreland Basin Development*; Abstract of the 23rd Earth Science Congress; Geological Society of South Africa: Cape Town, South Africa, 1990; pp. 250–253.

32. Kukla, P.A.; Stanistreet, I.G. Record of the Damara Khomas Hochland accretionary prism in central Namibia: Refutation of an "ensialic" origin of a Late Proterozoic orogenic belt. *Geology* **1991**, *19*, 473–476. [CrossRef]

33. Kukla, P.A. Tectonics and sedimentation of a late Proterozoic Damaran convergent continental margin, Khomas Hochland. *Memoir. Geol. Surv. Namib.* **1992**, *12*, 1–95.

34. Blanco, G.; Germs, G.J.B.; Rajesh, H.M.; Chemale, F., Jr.; Dussin, I.A.; Justino, D. Provenance and paleogeography of the Nama Group (Ediacaran to early Paleozoic, Namibia): Petrography, geochemistry and U-Pb detrital zircon geochronology. *Precambr. Res.* **2011**, *187*, 15–32. [CrossRef]

35. Nascimento, D.B.; Ribeiro, A.; Trouw, R.A.J.; Schmitt, R.S.; Passchier, C.W. Stratigraphy of the Neoproterozoic Damara Sequence in northwest Namibia: Slope to basin sub-marine mass-transport deposits and olistolith fields. *Precambr. Res* **2016**, *278*, 108–125. [CrossRef]

36. Porada, H. The Damara-Ribeira orogen of the Pan-African/Brasiliano cycle in Namibia (South West Africa) and Brazil as interpreted in terms of continental collision. *Tectonophysics* **1979**, *57*, 237–265. [CrossRef]

37. Miller, R.M. The Pan-African Damara orogen of south west Africa/Namibia. In *Evolution of the Damara Orogen of South West Africa/Namibia*; Miller, R.M., Ed.; The Geological Society of South Africa: Johannesburg, South Africa, 1983; Volume 11, pp. 431–515.

38. Prave, A.R. Tale of three cratons: Tectostratigraphic anatomy of the Damara Orogen in northwestern Namibia and the assembly of Gondwana. *Geology* **1996**, *24*, 1115–1118. [CrossRef]

39. Puhan, D. Reverse age relations of talc and tremolite deduced from reaction textures in metamorphosed siliceous dolomites of the southern Damara Orogen (Namibia). *Contrib. Mineral. Petrol.* **1988**, *98*, 24–27. [CrossRef]

40. Smith, D.A.M. The geology around the Khan and Swakop Rivers in South West Africa. *Memoir. Geol. Surv. S. Afr. S.W. Afr. Ser.* **1965**, *3*, 113.

41. Jacob, R.E.; Snowden, P.A.; Bunting, F.J.L. Geology and structural development of the tumas basement dome and its cover rocks. In *Evolution of the Damara Orogen of South West Africa/Namibia*; Milled, R.M., Ed.; Geological Society of South Africa, Special Publisher: Johannesburg, South Africa, 1983; Volume 11, pp. 157–172.

42. Kröner, A. Dome structures and basement reactivation in the Pan-African Damara belt of Namibia/South West Africa. In *Precambrian Tectonics Illustrated*; Kröner, A., Greiling, R.O., Eds.; Naegele and Obermiller: Stuttgart, Germany, 1984; pp. 191–206.

43. Oliver, G.J.H. Mid-crustal detachment and domes in the central zone of the Damara Orogen, Namibia. *J. Afr. Earth Sci.* **1994**, *19*, 331–344. [CrossRef]

44. Poli, L.C.; Oliver, G.J.H. Constrictional deformation in the Central Zone of the Damara Orogen, Namibia. *J. Afr. Earth Sci.* **2001**, *33*, 303–321. [CrossRef]

45. Alexander, F.M.; Kisters, L.; Smith, J.; Kathrin, N. Thrust-related dome structures in the Karibib district and the origin of orthogonal fabric domains in the south Central Zone of the Pan-African Damara belt, Namibia. *Precambri. Res* **2004**, *133*, 283–303. [CrossRef]

46. Jacob, R.E. Geology and Metamorphic Petrology of Part of the Damara Orogen along the Lower Swarkop River, South West Africa. Bulletin of the Precambrian res Unit. Ph.D. Thesis, University of Cape Town, Cape Town, South Africa, 1974.

47. Smith, E.F.; MacDonald, F.A.; Crowley, J.L.; Hodgin, E.B.; Schrag, D.P. Tectonostratigraphic evolution of the *c.* 780–730 Ma Beck Spring Dolomite: Basin Formation in the core of Rodinia. *Geol. Soc. Lond.* **2016**, *424*, 213–239. [CrossRef]

48. Marian, M.L. Sedimentology of the Beck Spring Dolomite, Eastern Mojave Desert, California. Master's Thesis, University of Southern California, Los Angeles, CA, USA, 1979.

49. Corsetti, F.A.; Kaufman, A.J. Stratigraphic investigations of carbon isotope anomalies and Neoproterozoic ice ages in Death Valley, California. *Geol. Soc. Am. Bull.* **2003**, *115*, 916–932. [CrossRef]

50. Whitney, D.L.; Evans, B.W. Abbreviations for names of rock-forming minerals. *Am. Mineral.* **2010**, *95*, 185–187. [CrossRef]

51. Hubbard, C.R.; Snyder, R.L. RIR-measurement and use in quantitative XRD. *Powder Diffr.* **1988**, *3*, 74–77.

52. RRUFF™ Project Database. Available online: http://rruff.info/ (accessed on 20 February 2019).

53. Maire, E.; Withers, P.J. Quantitative X-ray tomography. *Int. Mater. Rev.* **2014**, *59*, 1–43. [CrossRef]

54. Paganin, D.; Mayo, S.C.; Gureyev, T.E.; Miller, P.R.; Wilkins, S.W. Simultaneous phase and amplitude extraction from a single defocused image of a homogeneous object. *J. Microsc.* **2002**. [CrossRef]

55. Leake, B.E.; Woolley, A.R.; Arps, C.E.; Birch, W.D.; Gilbert, M.C.; Grice, J.D.; Hawthorne, F.C.; Kato, A.; Kisch, H.J.; Krivovichev, V.G.; et al. Nomenclature of amphiboles: Report of the subcommittee on amphiboles of the international mineralogical association, commission on new minerals and mineral names. *Can. Mineral.* **1997**, *35*, 219–246.

56. Witek, J.; Psiuk, B.; Naziemiec, Z.; Kusiorowski, R. Obtaining an artificial aggregate from cement-asbestos waste by the melting technique in an arc-resistance furnace. *Fibers* **2019**, *7*, 10. [CrossRef]

57. Witek, J.; Kusiorowski, R. Neutralization of cement-asbestos waste by melting in an arc-resistance furnace. *Waste Manag.* **2017**, *69*, 336–345. [CrossRef]

58. Spasiano, D.; Pirozzi, F. Treatments of asbestos containing wastes. *J. Environ. Manag.* **2017**, *204*, 82–91. [CrossRef]

fibers

MDPI

Article

Multi-Analytical Approach for Asbestos Minerals and Their Non-Asbestiform Analogues: Inferences from Host Rock Textural Constraints

Gaia Maria Militello [1,*], Andrea Bloise [2], Laura Gaggero [1], Gabriele Lanzafame [3] and Rosalda Punturo [4]

1   Department of Earth, Environment and Life Sciences – DISTAV, University of Genoa, Corso Europa 26, I-16132 Genoa, Italy; gaggero@dipteris.unige.it
2   Department of Biology, Ecology and Earth Sciences, University of Calabria, Via Pietro Bucci, I-87036 Rende, CS, Italy; andrea.bloise@unical.it
3   Elettra—Sincrotrone Trieste S.C.p.A., I-34149 Trieste, Italy; gabriele.lanzafame@elettra.eu
4   Department of Biological, Geological and Environmental Sciences, University of Catania, Corso Italia 55, I-95129 Catania, CT, Italy; punturo@unict.it
*   Correspondence: gaiamaria.militello@edu.unige.it; Tel.: +39-010-3538301

Received: 28 February 2019; Accepted: 3 May 2019; Published: 10 May 2019

Abstract: Asbestos is a hazardous mineral, as well as a common and well-known issue worldwide. However, amphiboles equal in composition but not in morphology, as well as the fibrous antigorite and lizardite, are not classified as asbestos even if more common than other forms of the mineral. Still, their potential hazardous properties requires further exploration. The proposed multi-instrumental approach focuses on the influence of textural constraints on the subsequent origin of asbestiform products in massive rock. This aspect has a significant effect on diagnostic policies addressing environmental monitoring and the clinical perspective. Concerning minerals that are chemically and geometrically (length > 5 μm, width < 3 μm and length:diameter > 3:1) but not morphologically analogous to regulated asbestos, the debate about their potential hazardous properties is open and ongoing. Therefore, a selection of various lithotypes featuring the challenging identification of fibrous phases with critical counting dimensions was investigated; this selection consisted of two serpentinites, one metabasalt and one pyroxenite. The analytical protocol included optical microscopy (OM), scanning and transmission electron microscopy combined with energy dispersive spectrometry (SEM/EDS; TEM/EDS), micro-Raman spectroscopy and synchrotron radiation X-ray microtomography (SR X-ray μCT). The latter is an original non-destructive approach that allows the observation of the fiber arrangement in a three-dimensional space, avoiding morphological influence as a result of comminution.

Keywords: texture; naturally occurring asbestos; morphology; cleavage fragments; asbestiform and fibrous minerals; environmental monitoring

1. Introduction

The term asbestos is a generic term comprising some natural minerals represented by hydrated silicates that are easily separable in thin, flexible fibers, resistant to traction and heat and almost chemically inert. The minerals defined as asbestos include the asbestiform varieties of minerals belonging to the amphibole group, such as riebeckite (under the commercial name of crocidolite), cummingtonite-grunerite series (under the commercial name of amosite), tremolite, actinolite and anthophyllite, as well as minerals belonging to the serpentine group (chrysotile). In detail, the term asbestiform refers to a specific type of mineral fibrosity, with high tensile strength and/or flexibility [1].

Therefore, in national and international normative definitions worldwide, minerals defined as asbestos include only fibrous and asbestiform varieties of the serpentine and amphibole groups, i.e., only those varieties that possess high tensile strength or flexibility. Meanwhile, the prismatic varieties of amphiboles, despite the fact that they have the same chemical composition, are not classified as asbestos. In terms of geometrical ratios, they would be classified as fibers (length >5 µm, diameter <3 µm and length/diameter ratio >3:1), but they are not asbestiform minerals; however, a clear relationship between their morphology and toxicity is still undemonstrated.

Nevertheless, even these geometric relationships are not univocally constrained; in fact, different counting criteria have been adopted for the assessment of asbestos risk and its identification [2]. Furthermore, it is difficult to discriminate the asbestos amphibole from prismatic crystals or cleavage fragments, especially if they are associated with one another.

Although it has been banned since 1992 in Italy and in many other countries worldwide [3], the management of asbestos is ruled by quite obsolete legislation and does not take into account many parameters for correct phase classification. This leads to problems of compatibility and reproducibility between different laboratories. Another concern is the classification of non-asbestos classified fibers [4] whose potential hazard is not fully investigated, such as prismatic and acicular habits or cleavage fragments of amphiboles and other elongated mineral particles (EMP) that are even more common [5]. Research and debate are in progress about the composition and morphology of the various types of asbestos and the related fibrous minerals not classified as asbestos, such as erionite and fibrous antigorite [6,7], which are similar in morphology to the polymorph chrysotile of asbestos and not always distinguishable. This morphology is regulated as asbestos only in the New Caledonia legislations [8]. As a consequence, quantitative determinations can yield high variance even for similar conditions or naturally occurring asbestos (NOA) types [9,10], preventing a risk assessment based on univocal data.

Therefore, the distinction between asbestos and non-asbestiform analogues is flawed from both a scientific and a regulatory standpoint [11–13]. Moreover, the habit at the macroscale does not always correspond to that observed at the microscale (i.e., not repeated with fractal geometries) [14]. In most laboratories phase-contrast light microscopy (PCM) or transmission electron microscopy with energy-dispersive X-ray analysis and selected-area electron diffraction (TEM/EDS/SAED) are traditionally used to characterize and evaluate the fiber concentrations in NOA aggregates [15,16]. TEM/EDS/SAED with a higher magnification and the possibility of mineral phase speciation allows the identification of chemical and crystallographic structure [17,18]. For example, in many works, the diameter of the chrysotile as well as those of amphiboles fibers were more accurately measured by TEM micrographs [18]. However, when the elongated particles are very thin, even with TEM it may not be possible to adequately differentiate asbestiform fibers from prismatic crystals or cleavage fragments. Moreover, in recent years other analytical techniques (e.g., µ-Raman, thermal analysis) have been able to better discriminate and quantify the asbestos mineral occurrence in rocks [19–21].

In this context, the object of this work is to compare, at different observation scales, by a multi-instrumental approach, e.g., [22], the morphological features and the influence of textural constraints in massive samples which determine the origin of fibrous and asbestiform or fibrous but not asbestiform products, or rather the EMP of amphibole and serpentine groups. These EMP therefore comprise cleavage fragments of minerals belonging by composition, but not necessarily by morphology, to the group of asbestos minerals.

Moreover, a new application of the synchrotron radiation X-ray microtomography (SR X-ray µCT), not yet used before for asbestos detection, was adopted. This semi-destructive technique was chosen because it can help to better observe the arrangement of the fibers in the three dimensions and facilitate the comparative description of cleavage fragments. Specially, if subjected to comminution, the preliminary step for the preparation of the samples according to the methods is regulated by law [1,23]. In order to understand whether the counting dimensions (length >5 µm, diameter <3 µm, length: diameter >3:1) have a bearing on the quantification, as a result of the comminution from an acicular or a fibrous crystal, five samples of metamorphic lithotypes containing serpentine and/or amphibole asbestos were addressed in the following multi-analytical investigations.

2. Materials and Methods

Dataset and Geological Provenance of Samples

The dataset is represented by two serpentinites (S1 and S2), one metabasalt (S3) and one metasomatized pyroxenite (S4).

Sample S1 (Piedmont, Italy) is a serpentinized peridotite cut by a fibrous/acicular lizardite-filled vein (Figure 1a). Sample S2 (Aosta Valley, Italy) is a serpentinite cut by a vein of dolomite, acicular and more or less fibrous tremolite and rare chrysotile (Figure 1b). Sample S3 (Piedmont, Italy) is a metabasalt with an intersertal texture where skeletal albite and acicular/prismatic and occasionally fibrous actinolite occur. The rock is cut by a plagiogranite vein formed mainly by primary plagioclase, secondary calcite and actinolite (Figure 1c). Sample S4 (Gauteng, South Africa) is a pyroxenite cut by millimeter-thick veins of the fibrous/acicular form of the actinolite-tremolite series associated with talc; the protolith is a metasomatized metagranitoid (Figure 1d).

Figure 1. Close up photographs of the analyzed samples: (**a**) serpentinized peridotite; (**b**) serpentinite with a tremolite vein; (**c**) metabasalt with a plagiogranite vein; (**d**) pyroxenite cut by a talc and tremolite-filled vein.

These samples were chosen as lithotypes with different petrological evolutions. They were also chosen because the characterization and identification of the fibrous phases is challenging. Samples were analyzed by optical microscopy (OM) (ZEISS, Thornwood, NY, USA), scanning and transmission electron microscopy combined with energy dispersive spectrometry (SEM/EDS and TEM/EDS) (SEM/EDS: TESCAN, Brno, Czech Republic; TEM/EDS: Jeol, Tokyo, Japan), μ-Raman (HORIBA, Longjumeau, France) and synchrotron radiation X-ray microtomography (SR X-ray μCT) (Hamamatsu, Hamamatsu City, Japan). For the latter technique, S1 underwent microtomographic analysis but was subsequently discarded due to the high and almost equal density of the mineralogical phases, and the consequent low-density contrast and lack of voids or fractures, which determined an almost total absorption of the X-rays and prevented the collection of significant information.

3. Analytical Methods

Many of the naturally occurring asbestos (NOA) types which are either fibrous or asbestiform and therefore regulated, show very similar optical characteristics. For this reason, their recognition requires in most cases the combined use of various analytical techniques, such as optical microscopy (OM), scanning electron microscopy with microanalysis (SEM/EDS), transmission electron microscopy with microanalysis (TEM/EDS), Fourier transform infrared spectroscopy, Raman spectroscopy and X-ray powders diffractometry. In our work we also added the use of an unconventional technique for the detection and characterization of NOA, namely synchrotron radiation microtomography (SR-μCT).

Optical microscopy (OM) is the fastest of these techniques for the recognition of different phases, and possibly of their morphology. Moreover, it provides an excellent indication of microstructural relationships within a rock. Most determinations were carried out with a Zeiss Axiolab Microscope with Polarized Light, located at the Department of Biological, Geological and Environmental Sciences of the University of Catania, Italy.

Scanning electron microscopy (SEM) makes possible morphological investigations under higher magnifications (up to 10,000×) with coupled in situ microanalysis to obtain compositional information. Sample preparation and qualitative determination were accomplished at the Electron Microscopy Laboratory of the Earth Environment and Life Sciences Department (DISTAV), University of Genoa, Italy, using a scanning electron microscope (TESCAN 3 XML) (TESCAN, Brno, Czech Republic). The work parameters were: 2000× of magnification (Mag) and 20 kV of accelerating voltage (HV). The elemental analysis of the minerals was carried out by energy dispersive X-ray spectroscopy (Oxford Instruments, AZtec 2.4) (TESCAN, Brno, Czech Republic). According with dimensions and chemical composition, minerals were detected and considered as asbestos fibers. Other EMP were analyzed to characterize them.

Micro-Raman spectroscopy allows the rapid identification of fibrous phases, providing information on the different vibrational modes of molecules. In particular, this technique has the advantage of being able to use the sample as is by positioning it on the stage of the microscope, or the same thin sections prepared for the observations in optical microscopy can be used. Micro-Raman scattering measurements were conducted at the Earth Environment and Life Sciences Department (DISTAV), University of Genoa, Italy, with a Horiba Jobin-Yvon Explora_Plus single monochromator spectrometer (HORIBA, Longjumeau, France) (with a grating of 2400 groove/mm) equipped with an Olympus BX41 microscope (HORIBA, Longjumeau, France). Raman spectra were excited by the 532 nm line. The spectrometer was calibrated to the silicon Raman peak at 520.5 cm^{-1}. The spectral resolution was ~2 cm^{-1} and the instrumental accuracy in determining the peak positions was ~0.56 cm^{-1}. Raman spectra were collected in the spectral ranges of 100–1100 cm^{-1} and 3000–3800 cm^{-1} for 15 s, averaging over 10 accumulations. In this work, this technique was used for the serpentine mineral identification of sample S1. Three analysis points were performed on three different parts of the sample: in the host rock, in the vein/rock interface and finally in the vein.

The transmission electron microscopy (TEM) associated with the microanalysis system implements the previous techniques, reaching atomic resolutions that allow the recognition of the different minerals of the serpentine sample and of all types of amphiboles. However, the technique addresses powders and therefore does not preserve the morphological and microstructural information of the sample. Finally, the preparation of the sample requires expert handling. Transmission electron microscopy (TEM), installed at the Department of Biology, Ecology and Earth Sciences, University of Calabria (Cosenza, Italy), was performed using a Jeol JEM 1400 Plus (Tokyo, Japan) working at 120 kV, equipped with a double tilt holder to check the morphology of the samples and to obtain structural data by selected area electron diffraction (SAED). Moreover, energy dispersive X-ray spectrometry (EDS) performed using the Jeol allowed us to obtain analytical electron microanalyses (AEM). In order to discriminate the asbestos fibers chemically, a three-point analysis were carried out on each single fiber. For TEM investigations, a small aliquot of the sample was ground using an agate pestle and mortar in isopropyl alcohol and then sonicated; this powder was then deposited onto a copper mesh grid coated with 200 Å carbon film. With TEM/EDS, only fibrous morphologies were investigated.

Synchrotron Radiation Microtomography (SR-μCT) Measurements

The three-dimensional study of three samples was performed at the SYRMEP beamline of the Elettra synchrotron laboratory (Trieste, Italy) by using high-resolution SR-μCT in phase-contrast mode [24].

The selected samples were cut in the form of a parallelepiped with a square base with dimensions of 2×0.5 cm^2 and illuminated by a polychromatic X-ray beam (white beam configuration) in transmission geometry. The contribution of low energies in the beam spectrum was suppressed by applying 1 mm Si + 1 mm Al filters. The SR-μCT was performed with a fixed sample-to-detector distance of 200 mm and collected, for each sample, 1800 projections over a total scan angle of 180° with an exposure time per projection of 2 s. The employed detector was a 16 bit, air-cooled, sCMOS camera (Hamamatsu C11440 22C) (Hamamatsu, Hamamatsu City, Japan) with a 2048 × 2048 pixel chip and an effective pixel size set at 1.95^2 μm^2, yielding a maximum field of view of ca. 3.2^2 mm^2. Since the lateral size of the samples was larger than the field of view of the detector, microtomographic scans were acquired in region-of-interest mode [25].

The 2D tomographic slices were reconstructed using the Syrmep Tomo Project (STP) house software suite [26], which allowed the application of different filters in order to reduce ring artefacts caused by detector inhomogeneity [27]. A single-distance phase-retrieval algorithm [28] based on the transport of intensity equation (TIE) was applied to the sample projections to improve the consistency of the morphological analysis. Combining phase-retrieval and filtered back-projection algorithms [29] allowed obtaining the 3D distribution of the complex refraction index of the imaged samples. This process reduces the edge-enhancement effect at sample borders, at the same time preserving the morphology of the smallest features.

The obtained 3D volumes were then segmented by manual thresholding. Three-dimensional renderings were obtained by VGStudio Max 2.2 software (Volume Graphics, Heidelberg, Germany).

4. Results

Prior to SR X-ray μCT, mineralogical and microtextural observations were carried out at the microscale, followed by a compositional and morphological analysis of polished thin sections. Microtextural sites with higher concentrations in fibrous and/or prismatic-acicular varieties of NOA were selected to be addressed to the microtomography runs.

4.1. Microtexural Observations

4.1.1. Sample S1: Serpentinized Peridotite

Since serpentine minerals have very similar optical properties and frequently have sub-microscopic intergrowths, especially when they have a fibrous habit (e.g., in veins) [30], it was necessary to begin the characterization of this sample with a preliminary analysis determined by μ-Raman. From the point analyses compared with the spectra in [31], it emerged that the vein (analysis spot 1 shown in Figure 2b) is represented by lizardite (Figure 2a,b) and that the interface between the vein and the host rock (analysis spot 2 in Figure 2d) is filled with chrysotile (Figure 2c,d).

The identification of the polymorphs was carried out by comparing the low- and high-wavenumber regions. In detail, the region of low-wavenumber (150–1000 cm^{-1}) corresponds to the vibrational mode of the crystal lattice and the vibrations of Si–O$_4$; contrariwise, in the high-wavenumber region the stretching vibration behavior of the OH group (3500–3800 cm^{-1}) is displayed. The low-wavenumber region is sometimes ambiguous because a little modification of the chemical composition could modify the intensity of the peak as well as the vibrational wave number and its intensity. Therefore, for the distinction of the serpentine polymorphs it is absolutely fundamental for the visualization of the vibration of the OH group, which is very sensitive to the variation of the geometry of the layers [31].

From the literature, the peaks at 230, 390 and 690 cm^{-1} are similar in both chrysotile and lizardite but are slightly shifted in antigorite. In particular, the peak at 230 cm^{-1} is associated with the O–H–O vibration in which the O is that not bound to the tetrahedra SiO_4 and the H is the hydrogen outside the OH group [31]. The peak at 390 cm^{-1} represents the mode of SiO_4 v_5 (e) and the peak at 692 cm^{-1} corresponds to the symmetrical Si–O_b–Si stretching.

In our case, the lizardite (Figure 2a) has very intense peaks at 690, 384 and 231 cm^{-1}; however, the peaks at 3682 and 3704 cm^{-1} are discriminating (Figure 2b).

The chrysotile (Figure 2c) shows peaks at 230, 385 and 690 cm^{-1}, while the OH group shows the main peak at 3697 cm^{-1} (Figure 2d) with a small hump of lower intensity.

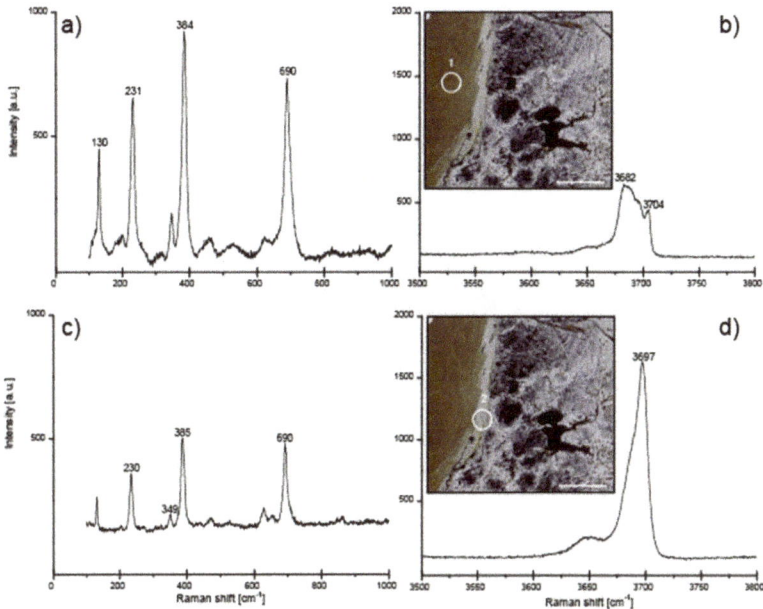

Figure 2. Raman spectra in the low- and high-wavenumber regions of lizardite ((**a,b**), respectively) and chrysotile ((**c,d**), respectively). Photomicrographs show the positions of the spots in which the spectra were acquired; scale bar: 0.2 mm.

Under OM, most of the host rock is made up of antigorite and it is very rich in Fe-oxides (mainly magnetite). Extensional veins with chrysotile/antigorite filling contain relics of pyroxenes (diopside) with prismatic habit and kinked cleavages. Within the diopside grains, small veins filled with fibrous talc were developed.

Therefore, all the three serpentine polymorphs occur in the sample. Although the in situ composition and speciation could be determined using μ-Raman, OM (Figure 3a) and SEM (Figure 3b,c), their habit could not be elucidated. The fiber morphology was better understood along fractures, discontinuities and close to the vein.

TEM investigations showed the four structural varieties of the serpentine group: chrysotile, lizardite, antigorite (both in platelets and fibers) and polygonal serpentine. Chrysotile consists of bundles of long and flexible fibrils with a hair-like appearance (Figure 3d). The chrysotile fibers are often curved and their length could be longer than 2 μm. TEM observations of single fibers showed cylindrical and rarely conical morphologies (Supplementary Materials Figure S1). The central empty core lies along the entire fibers length, sometimes with some interruptions (Supplementary Materials Figure S1), with outer and inner diameters of about 40 and 8 nm, respectively. In some fibers the outer walls of the chrysotile are very thin while the core is wide. In rare occasions, chrysotile grew with a proto-chrysotile morphology (Supplementary Materials Figure S1) [32]. The proto-chrysotile morphology is the precursor of the cylindrical morphology of chrysotile [18,33]. The polygonal serpentine fibers often have a diameter greater than 100 nm (Supplementary Materials Figure S1) as measured in the (100) cross-section. All of the other minerals mostly have a platy morphology and thus are identified as non-fibrous (e.g., lizardite, antigorite). The fibrous antigorite was rarely detected. In recent years, fibrous antigorite has been identified worldwide in some outcrops of serpentinite [17,18,34,35]. The bundle sides have a smooth and nearly constant diameter along the length.

Concerning the SR X-ray µCT investigation, it was not possible to proceed to the volume rendering of this sample due to the occurrence of Fe-oxides such as magnetite, where the high density absorbed most of the radiation, obscuring the other phases.

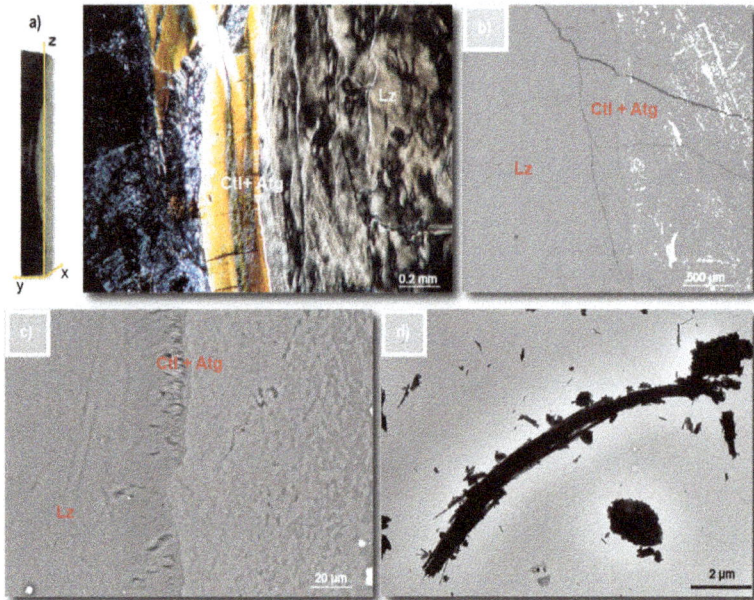

Figure 3. Sample 1 (S1). (**a**) Sample cut for tomographic scanning and photomicrography of the contact between the serpentinized peridotite and the lizardite vein; crossed polars. (**b**) SEM microphotograph (100× magnification) of serpentinized peridotite. (**c**) SEM microphotograph detail (2000× magnification) of the contact between the serpentinized peridotite and the lizardite vein (HV: 20 kV; Detector: Back-Scattered Electrons, BSE). (**d**) TEM image of a bundle of chrysotile fibers. Mineral name abbreviations are according to [36]. Atg: antigorite; Ctl: chrysotile; Lz: lizardite.

4.1.2. Sample S2: Serpentinite with Tremolite Vein

This sample mostly consists of a vein within serpentinite. The vein mainly constitutes dolomite and calcite (20 volume%), clinochlore, Fe-oxides (10 volume% with a grain size from 0.05 up to 1.2 mm) and tremolite (the rest of the sample). In addition, the sample is very rich in mega-crystals of chrysotile/fibrous antigorite affected by kink deformation. Tremolite, very variable in shape and size, is often associated with diopside in variable amounts along the vein cross-section. The grain size of tremolite is variable, from very fine (0.03 mm in length) up to a few millimeters in length. In addition, there are porphyroblasts that reach up to 4 mm in length. Moreover, the morphology of this phase becomes more fibrous due to the high aspect ratio of the particles. Tremolite occurs in an apparently prismatic/acicular habit with crystalline termination and well-defined edges or corners (Figure 4a); an asbestiform morphology is evidenced only close to voids (Figure 4b,c).

The apparent fibrous morphology at lower magnifications was sometimes confirmed, resulting from bladed grains similar to veritable asbestos fibers (Figure 4d). In fact, under TEM, tremolite shows the classical strain-shaped morphology with parallel sides and regular termination (Supplementary Materials Figure S2), while sometime the longitudinal splitting of the fibers parallel to (110) cleavage surfaces into thinner fibrils is observed (Supplementary Materials Figure S2). The fibers length is generally >6 μm with a width of <0.2 μm. We excluded the possibility of these being cleavage fragments of bladed prismatic tremolite because in that case the crystals would have shown irregular sides and blunt edges; as a rule, the cleavage fragments produced from the prismatic habits of the non-fibrous analogues of asbestos minerals are not veritable mineralogical fibers [37]. However, in the same samples cleavage fragments of tremolite were also detected (Supplementary Materials Figure S2). Antigorite and lizardite were also detected, although fibrous antigorite was always found in a low amount.

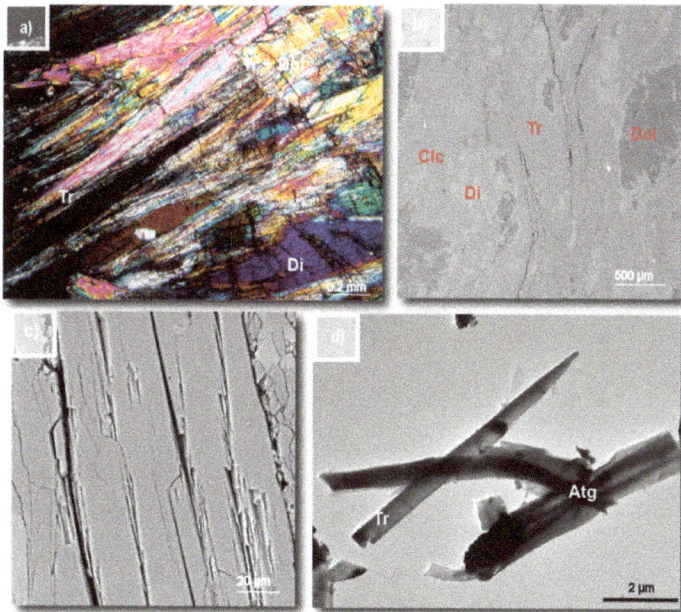

Figure 4. Sample 2 (S2). (**a**) Photomicrograph of the tremolite vein in the serpentinite; crossed polars. (**b**) SEM image (100× magnification) of the vein of the serpentinite. (**c**) SEM image detail (2000× magnification) of the acicular/fibrous tremolite (HV: 20 kV; Det: BSE). (**d**) TEM image of tremolite + fibrous antigorite. Mineral name abbreviations are according to [36]. Atg: antigorite; Clc: clinochlore; Di: diopside; Do: dolomite; Tr: tremolite.

Moreover, sample S2 was cut for tomographic scanning (Figure 5a). Starting from single slices of tomography (Figure 5b), this technique could obtain images of the 3D arrangement and the geometric relationship between the fiber-bearing and the massive sectors of the studied rock (Figure 5c).

In particular, by focusing on the 3D rendering (Figure 5c), a diopside- and tremolite-rich vein was evidenced, in which the tremolite fibers are stretched (dashes); around there, dolomite and clinochlore are in contact. Under SEM it was not possible to clearly correlate such an arrangement on the polished section, complementary to the phase.

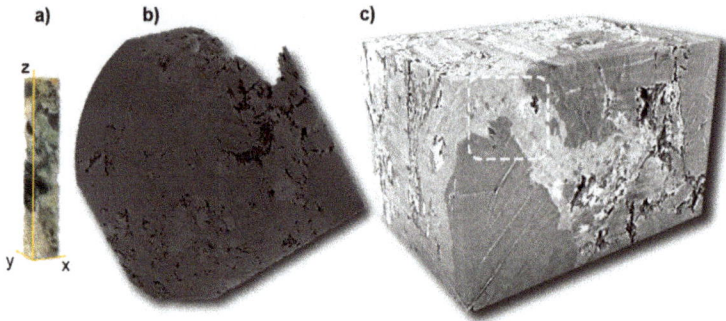

Figure 5. Sample 2 (S2). (**a**) Sample cut for tomographic scanning; (**b**) example of one single slice generated (1800 scans in total); (**c**) volume rendering (7.7 mm^3) obtained by synchrotron radiation X-ray microtomography (rendering performed using the commercial software VGStudio Max 2.0 (Volume Graphics, Heidelberg, Germany)).

4.1.3. Sample S3: Metabasalt with Plagiogranite Vein

This sample is a metabasalt with ophitic texture cut by a plagiogranite vein. Quartz, albite, and secondary calcite are the vein filling phases. Acicular or fibrous actinolite varieties occur either at the point of contact with the vein or are scattered in the basalt groundmass, which is rich in skeletal plagioclases and acicular actinolite. Titanite is the most common accessory mineral phase.

Amphiboles, sometimes as fibril bundles, are generally included in plagioclase. The very fine grain size inhibits the ability to determine by SEM whether the amphiboles are made of fibrils or cleavage fragments of prismatic amphiboles. Acicular and very elongated actinolite crystals occur at the contact between metabasalt and the plagiogranite vein (Figure 6a). The acicular morphology tends to thin out at the tips, which are often needle-like, with a generally moderate aspect ratio (10:1). Sometimes fibrils and acicular crystals coexist (Figure 6b,c) in a tight envelope, as is particular evident in 3D images (Figure 7).

TEM investigations showed amphibole fibers with a ragged surface and irregular sides, typical of the cleavage fragment origin (Figure 6d; Supplementary Materials Figure S3). According to the crystals/chemical data obtained by EDS/TEM, the elongated particles of amphiboles were classified as actinolite, since they had a value of Si ranging from 7.90 to 7.99 atoms per formula unit (a.p.f.u.) and a Mg/(Mg+Fe^{2+}) value >0.9 a.p.f.u. [38]. Furthermore, low concentrations of very short chrysotile fibers were also identified (Supplementary Materials Figure S3).

In the sample slice (Figure 7b), the actinolite overgrows the plagiogranite and the plagioclase inside the metabasalt. The 3D reconstruction reveals the texture and the rugged contact between the metabasalt and the plagiogranite (Figure 7c).

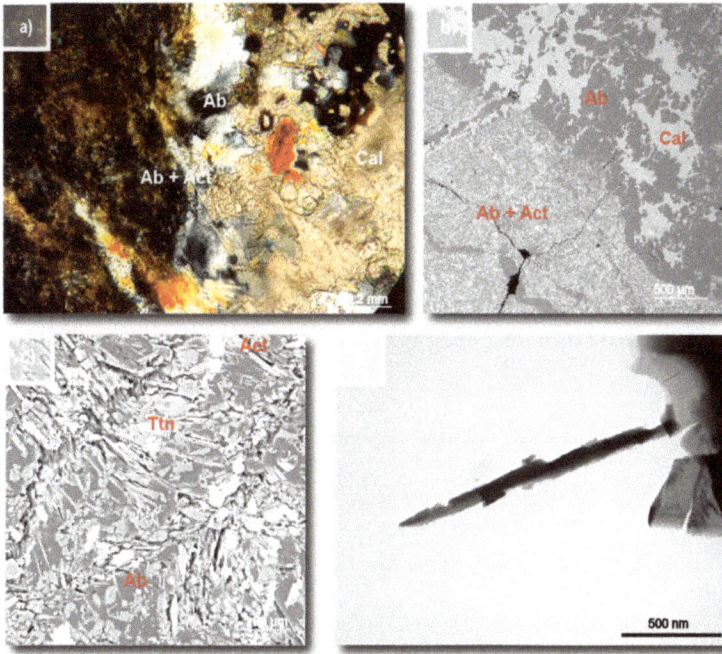

Figure 6. Sample 3 (S3). (**a**) Photomicrograph of the metabasalt and plagiogranite vein; crossed polars. (**b**) SEM image (100× magnification) of the metabasalt and plagiogranite vein. (**c**) SEM image (2000× magnification) detail of the acicular actinolite (HV: 20 kV; Det: BSE). (**d**) TEM image of acicular actinolite. Mineral name abbreviations are according to [36]. Ab: albite; Act: actinolite; Cal: calcite; Ttn: titanite.

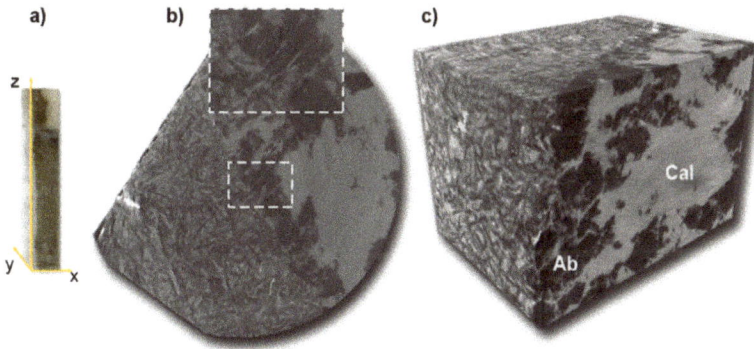

Figure 7. Sample 3 (S3). (**a**) Sample cut for tomographic scanning; (**b**) example of one single slice generated (1800 scans in total); (**c**) volume rendering (7.5 mm³) obtained by synchrotron radiation microtomography (rendering performed using the commercial software VGStudio Max 2.0). Mineral name abbreviations are according to [36]. Ab: albite; Cal: calcite.

4.1.4. Sample S4: Pyroxenite

This sample is a recrystallized cumulus rock and is mainly composed of granular clinopyroxene (diopside, ~80 volume%) with a mosaic texture as the orthocumulus phase, subhedral richterite (5–10 volume%) as an inter-cumulus phase and local cataclastic, rare orthopyroxene (3 volume%), tremolite (2 volume%), interstitial mica (phlogopite), quartz and apatite as accessory minerals and finally trace spinel (overall ~5 volume%).

The diopside is cut by talc micro-veins, and sometimes associated with titanite, ilmenite and tremolite. Part of the sample is a large vein, filled with tremolite, talc, calcite and chrysotile/antigorite with a finer grain size than clinopyroxene. Both fibrous and acicular habits are present. The grain size of the cumulus part of the sample is characterized by clinopyroxene, which has a grain from 0.2 to 8.5 mm, richterite from 0.05 to 1.5 mm, orthopyroxene from 0.5 up to 1.5 mm and oxides with an average grain size of 0.1 mm. In the cataclastic band, tremolite has an average grain size from 0.04 mm up to 0.4 mm and richterite has an average grain size from 0.06 to 0.3 mm, whereas the orthopyroxene is a fine-grained unresolvable aggregate.

Prismatic and fine-grained fibrous amphiboles of the tremolite/actinolite series are evidenced in Figure 8a. In the tremolite and talc vein, fibers tend to develop (Figure 8b,c) into a 3D network (Figure 9). The richterite sometimes shows a fibrous habit. Therefore, although the fibers are asbestiform, they are not regulated asbestos due to their mineral composition.

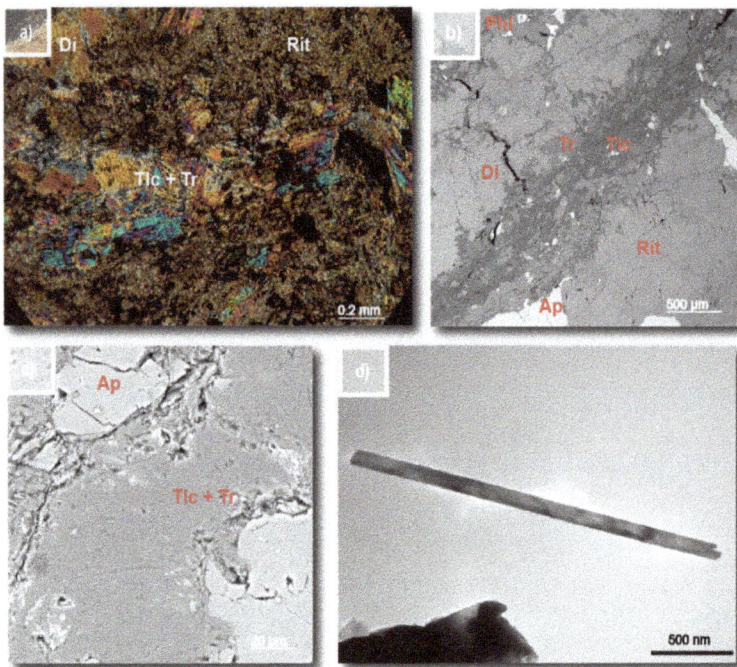

Figure 8. Sample 4 (S4). (**a**) Photomicrograph of the pyroxenite cut by a talc and tremolite vein; crossed polars. (**b**) SEM image (100× magnification) of pyroxenite cut by a talc and tremolite vein. (**c**) SEM image (2000× magnification) detail of the vein (HV: 20 kV; Det: BSE). (**d**) TEM image of fibrous actinolite. Mineral name abbreviations are according to [36] and references therein. Ap: apatite; Di: diopside; Phl: phlogopite; Rit: richterite; Tlc: talc; Tr: tremolite.

Fibers and elongated particles assemblage detected by TEM/EDS in the following samples were: asbestos actinolite, asbestos tremolite, cleavage fragment of tremolite and chrysotile. Asbestos actinolite exhibits a lath-shaped morphology (Figure 8d), as does asbestos tremolite (Supplementary Materials Figure S4). On the other hand, in the same samples, tremolite—classified as "non-asbestos" cleavage fragment—was observed (Supplementary Materials Figure S4). In fact, according to the literature data [5,39] the asbestiform habit is confirmed by considering the typical square terminations of the fibrils, whereas the sample shown in Supplementary Materials Figure S4 better accounts for an origin in the cleavage along the z axis of the tremolite, as suggested by the oblique tip of the fiber and the projection of the polygonal cross-section of the double chain building. Chrysotile, rarely detected in this sample, showed a classic cylindrical shape.

The elaboration of the synchrotron data showed that the veins rich in talc and tremolite cut the sample. This was also evident from the low-magnification SEM image (Figure 8b), and from the vein direction and texture shown in both the 2D axial slice and 3D rendering (Figure 9a–c).

Figure 9. Sample 4 (S4). (**a**) Sample cut for tomographic scanning; (**b**) example of one single slice generated (1800 scans in total); (**c**) volume rendering (7.4 mm^3) obtained by synchrotron radiation X-ray microtomography (rendering performed using the commercial software VGStudio Max 2.0). Mineral name abbreviations are according to [36]. Di: diopside; Rit: richterite; Tlc: talc; Tr: tremolite.

5. Discussion and Conclusions

The observations under OM allowed us to investigate the textures and the spatial relations between the mineralogical phases. The identification of asbestos minerals is easy under OM, but even at higher magnifications a fibrous habit has the same geometric parameters of prismatic/acicular crystals. Under SEM it is better characterize the aspect ratios and the textural relations between the various phases—often the habit is detectable, but discrimination among the serpentine polymorphs when embedded in the host rock is impossible.

TEM observation permitted the high-resolution discrimination for EMP and an effective distinction between regulated and non-regulated asbestos minerals. Therefore, combining all the techniques, the analyzed samples were found to be mainly composed of chrysotile and fibrous antigorite (S1); acicular associated with less fibrous tremolite and rare chrysotile (S2); acicular to prismatic actinolite (S3); and acicular/fibrous amphiboles of the actinolite-tremolite series (S4).

Preliminary results provided by SR X-ray μCT results showed that the serpentinized peridotite with the lizardite vein did not provide effective information, as the presence of magnetite (high density) caused the adsorption of most of the radiation, obscuring the other phases. Conversely, the other three samples exhibited good discrimination.

On the whole, synchrotron radiation X-ray microtomography is a technique that offers the possibility to observe and image the 3D morphology and the spatial relationship between the mineral phases that constituted the investigated samples. Furthermore, it is a semi-destructive technique which allows the analysis of samples under high magnification without grinding/milling and or the loss of their morphology. Nevertheless, it provides qualitative and volumetric but not quantitative information of the mineralogical phases; therefore, it proved a complementary technique to other conventional ones, useful to implement the research, but not exactly an effective activity for quantitative data restitution according to the legislative protocols in force for environmental monitoring [40,41].

It is noteworthy that none of the mentioned instruments (Table 1) is individually suitable for the unique and time-efficient identification of asbestos and non-asbestiform but still fibrous phases, particularly if these phases occur within massive rocks and are therefore not isolated.

Table 1. Summary of asbestos minerals and their fibrous but non-asbestiform analogues with their relative morphologies identified with the different analytical techniques used, respectively: Optical Microscopy (OM); Scanning Electron Microscopy combined with Energy Dispersive Spectrometry (SEM/EDS); Transmission Electron Microscopy combined with Energy Dispersive Spectrometry (TEM/EDS); Micro-Raman spectroscopy (μ-Raman); Synchrotron Radiation X-ray Microtomography (SR-μCT). Mineral name abbreviations are according to [36].

Sample	Lithotype	OM	SEM/EDS	TEM/EDS	μ-Raman	SR-μCT
S1	serpentinized peridotite	serpentine polymorphs	serpentine polymorphs	ctl + lz + platy and fibrous atg	ctl + atg + lz	-
S2	serpentinite with tremolite vein	prismatic/acicular tr + serpentine polymorphs	acicular/fibrous tr + serpentine polymorphs cleavage	tr (cleavage fragment and fibrous), platy and fibrous atg + lz	-	fibrous tr + di vein (~1, 7 mm lengths)
S3	metabasalt with plagiogranite vein	acicular/fibrous act	fragments and fibrous act	cleavage fragments of act + ctl	-	acicular act (~0, 117 mm length)
S4	pyroxenite	prismatic/fibrous tr-act series	prismatic/fibrous tr-act series	fibrous act + cleavage fragments/fibrous tr + ctl	-	tr + tlc vein (~0, 75 mm lengths)

If a complete, unique and comprehensive knowledge of asbestos-bearing rock is needed, it is important to plan and carry out a multi-analytical approach that takes into consideration the several aspects related to NOA occurrences within the various lithotypes.

Supplementary Materials: The following are available online at http://www.mdpi.com/2079-6439/7/5/42/s1, Figure S1. TEM images of: (a) a thin cylindrical chrysotile; (b) conical chrysotile with no-interrupted empty core; (c) poorly shaped proto-chrysotile indicated by black arrow; (d) cross-section of (100) polygonal serpentine. Figure S2. (a) TEM images of: tremolite asbestos as viewed perpendicular to the fiber axis; (b) flattened tremolite splitting longitudinally into thinner fibrils; (c) prismatic single crystals of tremolite (cleavage fragment). Figure S3. TEM image of single crystals of tremolite (cleavage fragment); note the irregular sides. Figure S4. (a) TEM images of: tremolite asbestos as viewed perpendicular to the fiber axis; (b) prismatic single crystals of tremolite (cleavage fragment); note the irregular sides. The wider end displays an initial split into two to three fibrils.

Author Contributions: Conceptualization, G.M.M. and L.G.; methodology, G.M.M., A.B. and G.L.; software, G.L.; validation, L.G., R.P. and A.B.; investigation, G.M.M., A.B., L.G., R.P. and G.L; resources, L.G. and R.P.; data curation, G.M.M.; writing—original draft preparation, G.M.M., L.G. and A.B.; writing—review and editing, G.M.M., A.B., L.G., R.P. and G.L.; supervision, L.G., R.P. and A.B.; funding acquisition, L.G., R.P. and A.B.

Funding: This research profited from the financial support of the "*Analisi delle proprietà microstrutturali, chimico-fisiche di materiali inorganici; determinazioni quantitative della composizione mineralogica di materiali naturali e delle proprietà tecniche dei materiali litici*" laboratory funds, DISTAV, University of Genoa, Italy. Moreover, part of this research was carried out with the financial support of "*Piano Triennale della Ricerca (2017–2020) and later*" (Deptartment of Biological, Geological and Environmental Sciences of the University of Catania, Italy.); scientific responsible: Rosalda Punturo. The work received financial support from the FFABR fund (by the Italian MIUR); scientific responsible: Andrea Bloise. Funding also came from the ELETTRA synchrotron laboratory (Trieste, Italy).

Acknowledgments: We gratefully acknowledge Elisa Sanguineti and Adrián Yus González for the support during the observation of the samples by SEM. We thank Nicola Campomenosi for help in acquisition and Simona Scrivano for the elaboration of the micro-Raman spectra. We thank Lucia Mancini, who was responsible for the SYRMEP beamline of the ELETTRA synchrotron laboratory (Trieste, Italy).

Conflicts of Interest: The authors declare no conflict of interest.

References

1. Italian Ministerial Decree No. 06/09/1994. All. 2B. Determinazione Quantitativa Delle Concentrazioni di Fibre di Amianto Aerodisperse in Ambienti Indoor Mediante Microscopia Elettronica a Scansione. Available online: http://www.earaonline.eu/wp-content/uploads/Decreto-Ministeriale-06-09-94.pdf (accessed on 1 April 2019).
2. Belardi, G.; Vignaroli, G.; Trapasso, F.; Pacella, A.; Passeri, D. Detecting asbestos fibres and cleavage fragments produced after mechanical tests on ophiolite rocks: clues for the asbestos hazard evaluation. *J. Mediterr. Earth Sci.* **2018**, *10*. Available online: http://jmes.it/index.php/jmes/article/view/111 (accessed on 1 April 2019).
3. Gualtieri, A.F. (Ed.) Introduction. In *Mineral Fibres: Crystal Chemistry, Chemical-Physical Properties, Biological Interaction and Toxicity*; European Mineralogical Union: London, UK, 2017; Volume 18, pp. 1–15.
4. Baumann, F.; Ambrosi, J.-P.; Carbone, M. Asbestos is not just asbestos: An unrecognised health hazard. *Lancet Oncol.* **2013**, *14*, 576–578. [CrossRef]
5. Belluso, E.; Cavallo, A.; Halterman, D. Crystal habit of mineral fibres. In *Mineral Fibres: Crystal Chemistry, Chemical-Physical Properties, Biological Interaction and Toxicity*; Gualtieri, A.F., Ed.; Mineralogical Society of Great Britain and Ireland, 2017; Volume 18, pp. 65–109. Available online: https://www.minersoc.org/emu-notes-18.html (accessed on 1 April 2019).
6. Ballirano, P.; Pacella, A.; Bloise, A.; Giordani, M.; Mattioli, M. Thermal Stability of Woolly Erionite-K and Considerations about the Heat-Induced Behaviour of the Erionite Group. *Minerals* **2018**, *8*, 28. [CrossRef]
7. Cardile, V.; Lombardo, L.; Belluso, E.; Panico, A.; Capella, S.; Balazy, M. Toxicity and carcinogenicity mechanisms of fibrous antigorite. *Int. J. Environ. Res. Public Health* **2007**, *4*, 1–9. [CrossRef]
8. Petriglieri, J.R.; Laporte-Magoni, C.; Salvioli-Mariani, E.; Gunkel-Grillon, P.; Tribaudino, M.; Bersani, D.; Lottici, P.P.; Mantovani, L.; Bursi Gandolfi, N. Fibrous minerals in New Caledonia: A comparison of different analytical strategies for environmental monitoring. Conference Paper: Congresso Congiunto SIMP-AIV-SoGeI-SGI. Geosciences: A Tool in a Changing World, Pisa, Italy, 3–6 September 2017; Available online: https://www.socgeol.it/files/download/pubblicazioni/Abstract%20Book/abstract_book_pisa_2017_doi.pdf (accessed on 1 April 2019).
9. Bloise, A.; Punturo, R.; Catalano, M.; Miriello, D.; Cirrincione, R. Naturally occurring asbestos (NOA) in rock and soil and relation with human activities: the monitoring example of selected sites in Calabria (southern Italy). *Ital. J. Geosci.* **2016**, *135*, 268–279. [CrossRef]
10. Bloise, A.; Belluso, E.; Critelli, T.; Catalano, M.; Apollaro, C.; Miriello, D.; Barrese, E. Amphibole asbestos and other fibrous minerals in the meta-basalt of the Gimigliano-Mount Reventino Unit (Calabria, south-Italy). *Rend. Online Soc. Geol. It.* **2012**, *21*, 847–848.
11. Harris, K.E.; Bunker, K.L.; Strohmeier, B.R.; Hoch, R.; Lee, R.J. Discovering the true morphology of amphibole minerals: complimentary TEM and FESEM characterization of particles in mixed mineral dust: In Modern Research and Educational Topics in Microscopy. *Modern Res. Educ. Top. Microsc.* **2007**, *3*, 643–650.
12. Punturo, R.; Cirrincione, R.; Pappalardo, G.; Mineo, S.; Fazio, E.; Bloise, A. Preliminary laboratory characterization of serpentinite rocks from Calabria (southern Italy) employed as stone material. *J. Mediterr. Earth Sci.* **2018**, *10*. [CrossRef]
13. Punturo, R.; Bloise, A.; Critelli, T.; Catalano, M.; Fazio, E.; Apollaro, C. Environmental natural implications related to asbestos occurrences in the ophiolites of the gimigliano-mount reventino unit (Calabria, southern Italy). *Int. J. Environ. Res.* **2015**, *9*, 405–418.
14. Gaggero, L.; Crispini, L.; Marescotti, P.; Malatesta, C. Solimano, 4-6 December. Structural and microstructural control on chrysotile distribution in serpentinites from eastern ligurian ophiolites. In *European Conference on Asbestos Risks and Management*; Rome, 2006; pp. 134–139. Available online: https://www.researchgate.net/profile/Pietro_Marescotti/publication/236862917_Structural_and_microstructural_control_on_chrysotile_distribution_in_serpentinites_from_eastern_ligurian_ophiolites/links/0c960519a489cde671000000/Structural-and-microstructural-control-on-chrysotile-distribution-in-serpentinites-from-eastern-ligurian-ophiolites.pdf (accessed on 1 April 2019).
15. National Institute for Occupational Safety and Health (NIOSH). Asbestos by TEM: 7402. issue 2. 15 August 1994. Available online: https://www.cdc.gov/niosh/docs/2003-154/pdfs/7402.pdf (accessed on 1 April 2019).
16. ASTM International, ASTM D7521-13, Standard Test Method for Determination of Asbestos in Soil, West Conshohocken, PA, 2013. Available online: https://standards.globalspec.com/std/10018512/astm-d7521 (accessed on 1 April 2019).

17. Bloise, A.; Critelli, T.; Catalano, M.; Apollaro, C.; Miriello, D.; Croce, A.; Barrese, E.; Liberi, F.; Piluso, E.; Rinaudo, C.; et al. Asbestos and other fibrous minerals contained in the serpentinites of the Gimigliano-Mount Reventino unit (Calabria, S-Italy). *Environ. Earth. Sci.* **2014**, *71*, 3773–3786. [CrossRef]

18. Bloise, A.; Catalano, M.; Critelli, T.; Apollaro, C.; Miriello, D. Naturally occurring asbestos: Potential for human exposure, San Severino Lucano (Basilicata, Southern Italy). *Environ. Earth Sci.* **2017**, *76*, 648. [CrossRef]

19. Rinaudo, C.; Belluso, E.; Gastaldi, D. Assessment of the use of Raman spectroscopy for the determination of amphibole asbestos. *Mineral. Mag.* **2004**, *68*, 455–465. [CrossRef]

20. Bloise, A.; Miriello, D. Multi-Analytical Approach for Identifying Asbestos Minerals in Situ. *Geoscience* **2018**, *8*, 133. [CrossRef]

21. Viti, C. Serpentine minerals discrimination by thermal analysis. *Am. Miner.* **2010**, *95*, 631–638. [CrossRef]

22. Compagnoni, R.; Groppo, C. Gli amianti in Val di Susa e le rocce che li contengono. *Rend. Soc. Geol. It.* **2006**, *3*. Nuova Serie, 21-28, 11 ff., 3 tabb. Available online: https://www.socgeol.it/files/download/workshop/05%20VS%20(21-28).pdf (accessed on 1 April 2019).

23. International Standardization Organization (ISO). ISO/DIS 22262-2, Bulk Materials, Part 2: Quantitative determination of Asbestos by Gravimetric and Microscopical Methods. 10.08.2009b. Available online: https://www.iso.org/standard/56773.html (accessed on 1 April 2019).

24. Cloetens, P.; Pateyron-Salome, M.; Buffière, J.Y.; Peix, G.; Baruchel, J.; Peyrin, F.; Schlenker, M. Observation of microstructure and damage in materials by phase sensitive radiography and tomography. *J. App. Phys.* **1997**, *81*. [CrossRef]

25. Maire, E.; Withers, P.J. Quantitative X-ray tomography. *Int. Mat. Rev.* **2014**, *59*, 1–43. [CrossRef]

26. Brun, F.; Massimi, L.; Fratini, M.; Dreossi, D.; Billé, F.; Accardo, A.; Pugliese, R.; Cedola, A. SYRMEP Tomo Project: a graphical user interface for customizing CT reconstruction workflows. *Adv. Struct. Chem. Imaging* **2017**, *3*. [CrossRef]

27. Brun, F.; Accardo, A.; Kourousias, G.; Dreossi, D.; Pugliese, R. Effective implementation of ring artifacts removal filters for synchrotron radiation microtomographic images. In Proceedings of the 8th International Symposium on Image and Signal Processing and Analysis (ISPA), Trieste, Italy, 4–6 September 2013; Ramponi, G., Lončarić, S., Carini, A., Egiazarian, K., Eds.; Available online: https://ieeexplore.ieee.org/document/6703823 (accessed on 1 April 2019).

28. Paganin, D.; Mayo, S.C.; Gureyev, T.E.; Miller, P.R.; Wilkins, S.W. Simultaneous phase and amplitude extraction from a single defocused image of a homogeneous object. *J. Microsc.* **2002**, *206*, 33–40. [CrossRef]

29. Herman, G.T. *Image Reconstruction from Projections: The Fundamentals of Computerized Tomography*, 1st ed.; Academic Press: New York, NY, USA, 1980; Available online: https://www.springer.com/us/book/9781852336172 (accessed on 1 April 2019).

30. Groppo, C.; Rinaudo, C.; Cairo, S.; Gastaldi, D.; Compagnoni, R. Micro-Raman spectroscopy for a quick and reliable identification of serpentine minerals from ultramafics. *Eur. J. Miner.* **2006**, *18*, 319–329. [CrossRef]

31. Petriglieri, J.R.; Salvioli-Mariani, E.; Mantovani, L.; Tribaudino, M.; Lottici, P.P.; Laporte-Magoni, C.; Bersani, D. Micro-Raman mapping of the polymorphs of serpentine. *J. Raman Spettrosc.* **2015**, *46*, 953–958. [CrossRef]

32. Bloise, A.; Kusiorowski, R.; Lassinantti Gualtieri, M.; Gualtieri, A.F. Thermal behavior of mineral fibers. In *Mineral Fibers: Crystal Chemistry, Chemical-Physical Properties, Biological Interaction and Toxicity*; Gualtier, A.F., Ed.; European Mineralogical Union: London, UK, 2017; Volume 18, pp. 215–252.

33. Yada, K.; Lishi, K. Growth and Microstructure of Synthetic Chrysotile. *Am. Miner.* **1977**, *62*, 958–965.

34. Belluso, E.; Compagnoni, R.; Ferraris, G. Occurrence of Asbestiform Minerals in the Serpentinites of the Piemonte Zone, Western Alps. 1995. Available online: https://iris.unito.it/handle/2318/23533#.XMxmpGixXIU (accessed on 1 April 2019).

35. Dogan, M.; Emri, S. Environmental health problems related to mineral dusts in Ankara and Eskisehir, Turkey. *Yerbilimleri* **2000**, *22*, 149–161.

36. Whitney, D.L.; Evans, B.W. Abbreviations for Names of Rock-Forming Minerals. *Am. Miner.* **2010**, *95*, 185–187. [CrossRef]

37. National Institute for Occupational Safety and Health (NIOSH). Asbestos Fibers and other Elongate Mineral Particles: State of the Science and Roadmap for Research. *Curr. Intell. Bull.* **2011**, *62*. Available online: https://www.cdc.gov/niosh/docs/2011-159/pdfs/2011-159.pdf (accessed on 1 April 2019).

38. Ballirano, P.; Bloise, A.; Gualtieri, A.F.; Lezzerini, M.; Pacella, A.; Perchiazzi, N.; Dogan, M.; Dogan, A.U. The Crystal Structure of Mineral Fibers. In *Mineral Fibers: Crystal Chemistry, Chemical-Physical Properties, Biological Interaction and Toxicity*; Gualtieri, A.F., Ed.; European Mineralogical Union: London, UK, 2017; Volume 18, pp. 17–53.
39. Van Orden, D.R.; Allison, K.A.; Lee, R.J. Differentiating Amphibole Asbestos from Non-Asbestos in a Complex Mineral Environment. *Indoor Built Environ.* **2008**, *17*, 58–68. [CrossRef]
40. Gaggero, L.; Sanguineti, E.; Yus González, A.; Militello, G.M.M.; Scuderi, A.; Parisi, G. Airborne asbestos fibres monitoring in tunnel excavation. *J. Environ. Manag.* **2017**, *196*, 583–593. [CrossRef]
41. Gaggero, L.; Crispini, L.; Isola, E.; Marescotti, P. Asbestos in natural and anthropic ophiolitic environments: A case study of geohazards related to the northern apennine ophiolites (Eastern Liguria, Italy). *Ofioliti* **2013**, *31*, 29–40.

![fibers logo] *fibers*

MDPI

Article

Assessment of Naturally Occurring Asbestos in the Area of Episcopia (Lucania, Southern Italy)

Andrea Bloise [1,*], Claudia Ricchiuti [2], Eugenia Giorno [3], Ilaria Fuoco [1], Patrizia Zumpano [1], Domenico Miriello [1], Carmine Apollaro [1], Alessandra Crispini [3], Rosanna De Rosa [1] and Rosalda Punturo [2]

[1] Department of Biology, Ecology and Earth Sciences, University of Calabria, Via Pietro Bucci, I-87036 Rende, Italy; ilaria.fuoco@unical.it (I.F.); patrizia85@gmail.com (P.Z.); miriello@unical.it (D.M.); apollaro@unical.it (C.A.); derosa@unical.it (R.D.R.)
[2] Department of Biological, Geological and Environmental Sciences, University of Catania, Corso Italia, 55, 95129 Catania CT, Italy; claudia.ricchiuti@unict.it (C.R.); punturo@unict.it (R.P.)
[3] MAT-InLAB-Department of Chemistry and Chemical Technologies, University of Calabria, 87036 Arcavacata di Rende-Cosenza, Italy; eugenia.giorno@unical.it (E.G.); a.crispini@unical.it (A.C.)
* Correspondence: andrea.bloise@unical.it; Tel.: +39-0984-493588

Received: 5 April 2019; Accepted: 14 May 2019; Published: 16 May 2019

Abstract: Over the last few years, the risk to human health related to asbestos fiber exposure has been widely demonstrated by many studies. Serpentinites are the main rocks associated with naturally occurring asbestos (NOA). In order to investigate the presence of NOA, a mineralogical study was conducted on eleven serpentinite samples collected nearby the village of Episcopia (Lucania, Southern Italy). Various analytical techniques such as X-ray powder diffraction (XRPD), scanning electron microscopy combined with energy dispersive spectrometry (SEM-EDS) and derivative thermogravimetry (DTG) were used to determine the occurrence of asbestos minerals and to make morphological observations. Results pointed out that all of the samples contain asbestos minerals (e.g., tremolite, actinolite and chrysotile). Moreover, it was observed that both natural processes and human activity may disturb NOA-bearing outcrops and provoke the formation of potentially inhalable airborne dust causing the release of asbestos fibers into the environment, thereby increasing the risk to human health. For this reason, our study aims to highlight the requirement of a natural asbestos survey and periodic update in the area.

Keywords: naturally occurring asbestos; serpentinites; polymorphs; health hazard

1. Introduction

Today, it is widely accepted in the scientific community that exposure to asbestos bring to the development of negative health issues. Indeed, malignant mesothelioma and lung cancers could be caused by the inhalation of asbestos fibers due to environmental exposure [1–4].

The silicate mineral habitus type may be fibrous or non-fibrous and, among the minerals which form the airborne particulate, the most hazardous ones display a fibrous-asbestiform crystal habitus [5]. The term Naturally occurring asbestos (NOA) refers to asbestos minerals contained in rocks and soils to distinguishing them from those contained in asbestos containing materials (ACM) [6–10]. Six fibrous silicate minerals belonging to the serpentine (i.e., chrysotile) and amphibole (i.e., tremolite, actinolite, anthophyllite, amosite, and crocidolite) mineral groups are defined as asbestos by law in Europe and in several countries worldwide [5]. However, many studies demonstrate that besides these six varieties, which are regulated as potential environmental pollutants by law, asbestiform minerals such as erionite, antigorite and fluoro-edenite could also be dangerous if respired by humans, leading to several respiratory diseases [11–17]. The issue is even more complicated as the US National Institute for Occupational Safety and Health (NIOSH) has lately proposed to extend the definition of asbestos to all elongated mineral particles (EMP) [18].

Chrysotile is one of the three principal serpentine polymorphs (chrysotile, lizardite, and antigorite), and it occurs with a fibrous habit [17]. Structurally, it is constituted by tetrahedral silica-oxygen groups (SiO_4) (T) connected to brucite-type $Mg(OH)_2$ octahedral sheets (O) by sharing of oxygen atoms, forming structures having the ideal formula $Mg_3Si_2O_5(OH)_4$ [17]. Because of the smaller dimension of the tetrahedral sheets to the corresponding octahedral ones, the connection of the TO layers results in a rolled papyrus-like microstructure which may compose a characteristic fibrous habit [18].

Amphiboles are built on double-chains of Si_4O_{11} groups linked to each other by a variety of cations, which may display a fibrous habit being structurally elongated in one preferred crystal direction [17]. The chemical composition of the amphibole group can be expressed by the general formula $AB_2C_5T_8O_{22}(OH)_2$, where A = Na and K; B = Na, Ca, Fe^{2+}, Mg, Mn^{2+}, Li and rarer ions of similar size; C = Fe^{2+}, Mg, Mn^{2+}, Li, Fe^{3+}, Cr^{3+}, Al, Mn^{3+}, Zr^{4+} and Ti^{4+}; T = Si, Al, and Ti^{4+}; and (OH^-) may be replaced by F^-, Cl^- and O^{2-} [17–21]. The A-site is in 10–12-fold coordination, while the B- and C-sites are octahedrally coordinated [17]. Amphiboles can be shown in monoclinic or orthorhombic crystalline system, and for both, modern nomenclature is based on the atomic proportions of the major elements assigned to the A, B, C and T structural sites, following the rules of Leake [19,20]. Among the amphibole group minerals, tremolite, actinolite and anthophyllite occur in both asbestiform and non-asbestiform habit types, whereas crocidolite and amosite occur only in the asbestiform habit [18].

Serpentinite rocks are widely outcropping in the Lucania region (Southern Italy) [22,23] and often they are removed from their natural place of origin to be used as ornamental stones and building materials due to their coloring and physic-mechanical properties [24]. However, asbestos tremolite/actinolite and/or chrysotile are detected in serpentinite outcrops of several urban centers of the region [25], including Episcopia. The release of asbestos fibers in the environment is the result of both natural weathering processes (e.g., erosion) and human activities (e.g., excavation or road construction), which may disturb NOA outcrops and provoke the formation of potentially inhalable airborne dust [6,21], causing one or more respiratory diseases that could occur after a long latency time (e.g., [1]). In particular, about 3000 people living around the study area, comprising Episcopia village and its hamlets (Figure 1), could be exposed to potential sources of airborne asbestos due to the wide distribution of outcrops where serpentinite is exploited.

Previous studies on serpentine rocks, carried out in the central and southern parts of the Basilicata region, highlighted that it is necessary to deepen public health studies in order to characterize and determine the position of NOA [22–26]. Moreover, a recent epidemiological study conducted in twelve villages located in this part of the region showed significant excesses in health problems that are NOA-correlated cases [27]. Particularly, in the geographic area located about 20 km from Episcopia, several mesothelioma cases were documented in which the etiological factor turned out to be exposure to asbestos minerals [28,29]. Therefore, local maps indicating areas where asbestos is present in outcropping rocks, as is established by Italian law (DM 18/03/2003), are crucial to avoid hazardous exposure to populations.

So far, a detailed mineralogical characterization of asbestos minerals present in the area of the Episcopia village is still lacking in the literature. In this context, aiming to point out the eventual presence of asbestos minerals within the serpentinite rocks cropping out in the surroundings of the Episcopia village (Figure 1), we collected eleven rock samples, studied them and crossed the data obtained from different analytical techniques (i.e., X-ray powder diffraction (XRPD), scanning electron microscopy combined with energy dispersive spectrometry (SEM-EDS) and derivative thermogravimetry (DTG)) for a detailed mineralogical characterization.

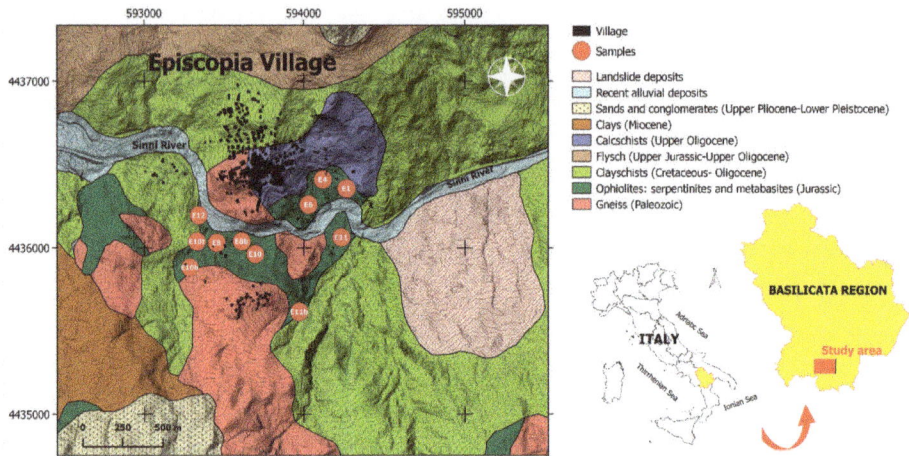

Figure 1. Simplified geological map of the Episcopia village and the study area location with the sample sites.

2. Materials and Methods

The Episcopia village (Figure 1) is located to the south-western part of the San Arcangelo basin in the Pollino National Park of the Basilicata region [30]. The pre-Pleistocene substrate (Episcopia-San Severino mélange) is discriminated by articulated tectonic slices overlapping that belong to different geological units and formations. Along the Sinni River, calc-schists and phyllites to the Unit of Frido crop out [31–35]. Over the Frido Unit, meta-ophiolites, made up of lenticular metabasite interbedded with and highly fractured serpentinites may occur (Figure 1) [32,36]. In a higher stratigraphic position, the Crete Nere and Saraceno formations and the Flysch of Albidona appear, which are non-metamorphic lithotypes referable to the North-Calabrian Unit [37].

With the aim to assess the presence of NOA in the area of the Episcopia village, 11 serpentinite rock samples were collected and characterized by X-ray powder diffraction (XRPD), scanning electron microscopy combined with energy dispersive spectrometry (SEM-EDS) and thermogravimetric analysis (TGA-DTG). Sample collection was conducted in the area surrounding the village, in particular the pieces were taken at the road cuts outside and within the village center and at dirt roads and slops in which serpentinites are well-exposed and display a vivid green color (Figure 2a,b).

X-ray powder diffraction (XRPD) data were obtained by X-ray diffraction acquired on a Bruker D2-Phaser (Bruker, Billerica, MA, USA) equipped with Cu Kα radiation ($\lambda = 1.5418$ Å) and a Lynxeye detector, at 30 kV and 10 mA, with a step size of 0.01° (2θ) and between 5 and 66° (2θ). The crystalline phases and semi-quantitative mineralogical composition of samples were obtained using EVA software (DIFFRACPlusEVA), which compares the experimental peaks with the 2005 PDF2 reference patterns. In the laboratory, the samples were recovered under a binocular microscope (20x, ZEISS, Thornwood, NY, USA) in order to choose representative fragments (i.e., more fibrous) to be studied by scanning electron microscopy. Scanning electron microscopy analysis combined with energy dispersive spectrometry (SEM-EDS) for the morphological observations was performed using an Environmental Scanning Electron Microscope FEI QUANTA 200 (Thermo Fisher Scientific, Waltham, MA, USA) equipped with an X-ray EDS suite comprising a Si/Li crystal detector model EDAX-GENESIS4000 (EDAX Inc., Mahwah, NJ, USA). For SEM examinations, a piece of each sample was firmed on an SEM stub utilizing double-sided conductive adhesive tape. In the present paper, the weight change was evaluated by thermogravimetric analysis (TGA: Netzsch STA 449 C Jupiter, Netzsch-Gerätebau GmbH, Selb, Germany). During thermogravimetric analysis (TGA) the samples were heated at a rate of 10 °C·min^{-1} in the 30–850 °C temperature range under an air flow of 30 mL·min^{-1}. About 20 mg of a sample were used for each run. Instrumental precision was checked by three repeated collections on a kaolinite reference sample revealing good reproducibility (instrumental theoretical T precision of ±1.2 °C). Derivative thermogravimetry (DTG) was obtained using the Netzsch Proteus thermal analysis software (Netzsch-Gerätebau GmbH, Selb, Germany). For X-ray powder diffraction and thermal analysis investigations, samples were mechanically milled using a Bleuler Rotary Mill (Sepor, Los Angeles, CA, USA) for 10 s at a speed of 900 revolutions per min (rpm).

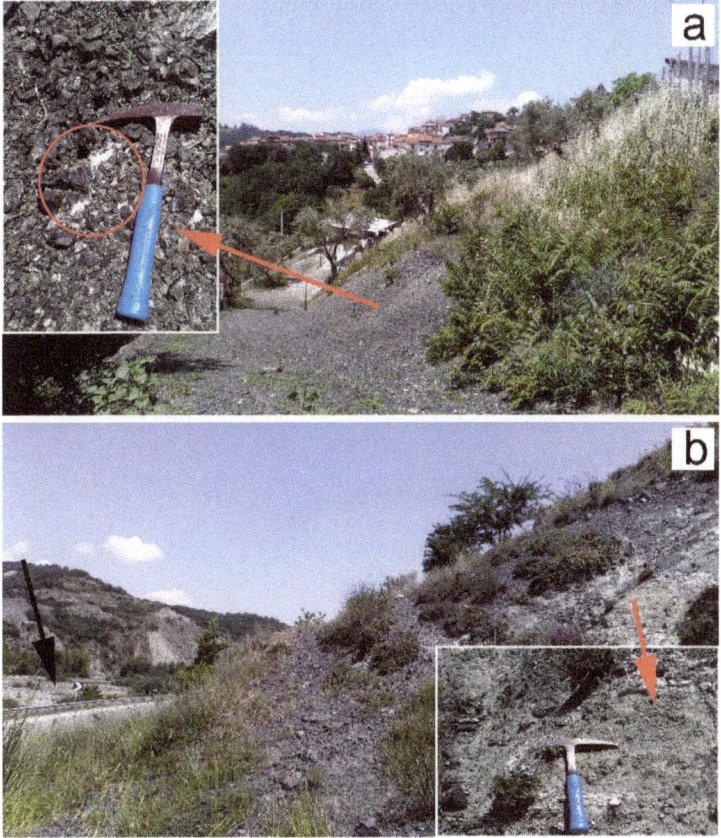

Figure 2. (**a**) Distant view of the Episcopia village, with appearance of serpentinite at the road cut near to the village (the red circle encloses white fibers of asbestos tremolite); and (**b**) pictures show vivid green serpentine at the road cut near to the village of Episcopia (highway 656 indicated by the black arrow). Inserts depict a zoomed-in area.

3. Results and Discussion

The field survey carried out nearby the Episcopia village (Figure 2) showed that serpentinite rocks appear to be characterized by a massive structure and by a dark-green color with widespread white parts consisting of fibrous minerals. Figure 3 shows the details of the studied samples at the mesoscopic scale, where it is possible to appreciate the appearance of white tremolite fibers (Figure 3a) and dark green serpentine (Figure 3b).

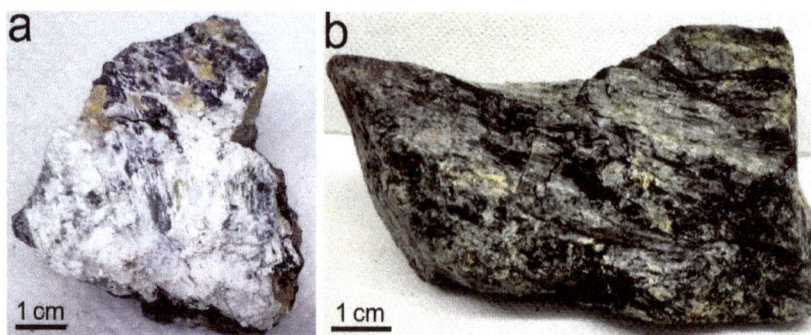

Figure 3. (a) White and silky tremolite on the surface of a serpentinite sample at the mesoscopic scale; and (b) the characteristic blazing surface of serpentinite, looking like a snake's skin.

The study conducted through various analytical techniques on eleven samples of serpentinite cropping out in the area of Episcopia village was finalized by determining the presence of NOA. Results of XRPD patterns showed that the investigated specimens are composed of serpentine minerals, chlorite, talc, tremolite, actinolite, willemseite and dolomite (Table 1). In particular, by the diffractograms interpretation the presence of serpentine minerals came out in all of the samples except for two, in which talc and tremolite (sample E10) and actinolite, willemseite, and dolomite (sample E10b) were the only phases detected. It is worth noting that the reflections diagnostic of the presence of asbestos amphiboles (i.e., tremolite/actinolite) were found in eight samples out of eleven (Table 1).

Table 1. Semi-quantitative mineralogical composition of samples in order of decreasing relative abundance, detected by X-ray powder diffraction (XRPD), scanning electron microscopy combined with energy dispersive spectrometry (SEM-EDS) and derivative thermogravimetry (DTG) analysis. Atg = antigorite, Lz = lizardite, Ctl = chrysotile, Act = asbestos actinolite, Tr = asbestos tremolite, PS = polygonal serpentine, Chl = chlorite, Will = willemseite, Dol = dolomite, and Tlc = talc. Mineral symbols after [38].

Sample	Site Description	Longitude East (WGS84)	Latitude North (WGS84)	Phases Detected Max ↔ Min
E1	Slope	594261	4436353	Ctl > Liz > Atg > Chl > Tlc > Tr
E4	Road cut	594118	4436407	Ctl > PS > Chl > Tlc > Tr
E6	Slope	594031	4436259	Chl > Tr > Liz > Ctl
E8	Dirt road	593457	4436029	Chl > Liz > Ctl
E8b	Dirt road	593611	4436038	Chl > Ctl > PS
E10	Slope	593696	4435960	Tlc > Tr
E10b	Road cut	593287	4435880	Wil > Dol > Act
E10t	Dirt road	593332	4436037	Tlc > Ctl > Atg > Act
E11	Road cut	594231	4436063	Tlc > Liz > Ctl,
E11b	Road cut	593973	4435615	Liz > Ctl > Tlc > Tr
E12	Road cut	593342	4436196	Tlc > Ctl > PS > Act

However, discrimination among the serpentine varieties (i.e., chrysotile, lizardite, antigorite, and polygonal serpentine) was not achievable by using only the X-ray powder diffraction method because diffraction peaks of the serpentine polymorphs overlap each other [7]. As reported in the literature, thermal analysis allowed for the discrimination among serpentine varieties [22,39]. Therefore, only samples in which serpentine minerals were detected by XRPD have been further investigated by thermal analysis. The correspondence between the maximum loss rates peaks and the serpentine minerals was defined in agreement with the literature data [40].

In particular, DTG curves showed the maximum peaks loss rate in the temperature range of 605–690 °C due to the chrysotile breakdown (Figure 4). The presence of DTG peaks in a temperature range of 705–731 °C were related to lizardite dehydroxylation, while antigorite occurred at higher temperatures (a 770 °C average value) than lizardite. Based on thermal analysis, chrysotile was identified in nine out of eleven analyzed samples, lizardite was detected in five samples, while antigorite and polygonal serpentine in two and three samples, respectively (Table 1).

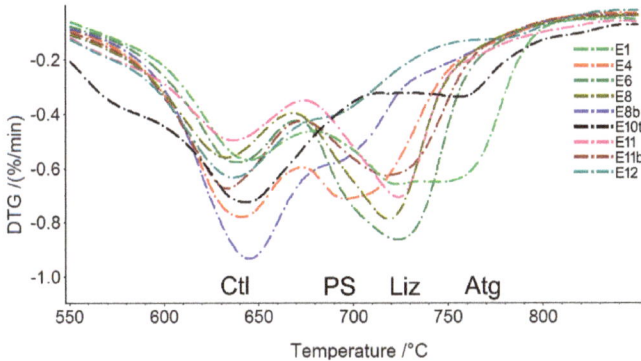

Figure 4. DTG curves for each sample in the temperature range of 500–850 °C. Endothermic peaks related to Ctl = chrysotile, PS = polygonal serpentine, Liz = lizardite, and Atg = antigorite. Mineral symbols after [38].

SEM observations highlighted that chrysotile is made up of either thin and flexible isolated fibril (Figure 5a) with a length longer than 6–8 μm or crystals arranged in bundles. In contrast, tremolite and actinolite appear straight and show a slender needle-like crystal habit with a length longer than 10 μm (Figure 5b). It is worth remembering that fibers are composed of many fibrils, which tend to split up along the fiber elongation axis [18]. This tendency leads to even smaller diameters, thus increasing the risks for human health related to the inhalation when they become airborne. Moreover, fibers having size matching with those of regulated asbestos (length >5 μm and an aspect ratio of 3) of both chrysotile and tremolite/actinolite have been detected in all samples, even if the length of both chrysotile and tremolite/actinolite fibers were sometimes shorter than the length established by law (Italian Legislative Decree 277/9).

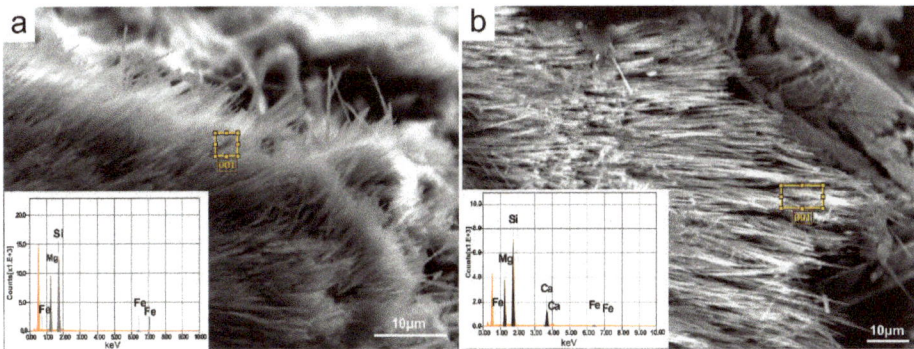

Figure 5. Scanning electron microscopy (SEM) images of asbestos. (**a**) Chrysotile sample E11; and (**b**) tremolite sample E4. Graphical inserts depict energy dispersive spectrometry (EDS) point analysis.

The use of the energy dispersive spectrometry (EDS) spot analysis (Figure 5a,b inserts) is essential for the correct identification of the chrysotile and tremolite/actinolite asbestos fibers. Chrysotile fibers show low amounts of Al, which mainly replaces the Mg in the octahedral sites. The EDS analyses revealed a low percent replacement of Mg for Fe occurs in the octahedral sites of chrysotile [22]. Regarding iron content, it ranges from a minimum of 3.51 wt % (sample E11) of FeO to 8.71 wt % (sample E12) with an mean value of 4.90 wt %. The presence of iron could play an significant function in the biological–mineral system interaction, increasing fiber toxicity, which has been unequivocally related to the effect of surface iron ions acting as catalytic sites generating free radicals and reactive oxygen species (ROS) [41].

The chemical composition of amphiboles detected by EDS was plotted in the diagram Si vs. $Mg/(Mg + Fe^{2+})$ [19], and graphically reported in Figure 6. Three samples (E10b, E10t, and E12) were plotted in the field of actinolite since their composition is: (i) an Si value to 7.94 a.p.f.u. (atoms per formula unit) and $Mg/(Mg + Fe^{2+})$ value to 0.87 (sample E10t); (ii) an Si value to 7.96 a.p.f.u. and $Mg/(Mg + Fe^{2+})$ value to 0.88 (sample E10b); and (iii) an Si value to 7.98 a.p.f.u. and $Mg/(Mg + Fe^{2+})$ value equal to 0.89. Five amphiboles were classified as tremolite (Table 1) since their chemical composition is: an Si range from 7.94 to 7.99 a.p.f.u. and a $Mg/(Mg + Fe^{2+})$ value >0.9.

The presence of iron in actinolite and tremolite could have a preeminent role in the biological–mineral system interaction. Indeed, it is worth pointing out that many researchers suggested that iron is a key component in asbestos toxicity [41–43]. Although some authors consider amphiboles (e.g., tremolite and crocidolite) to be more harmful than chrysotile to human health [43,44], all of the six asbestos minerals are assumed to be harmful. Therefore, in our opinion serpentinite samples containing asbestos are all potentially toxic for humans.

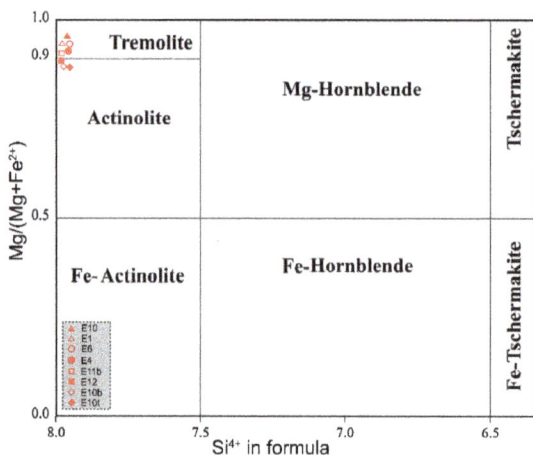

Figure 6. Amphibole classification diagram (after [19]).

Moreover, the analyses carried out by means of XRPD permitted also the identification of a nickelian-talc type named willemseite (Ni,Mg)$_3$Si$_4$O$_{10}$(OH)$_2$ never detected before in the study area (Table 1). Table 1 shows the list of the phases identified in each sample and chrysotile, tremolite and actinolite turn out to be common phases of the serpentinite rocks studied. Since the economy of the area is mainly based on sheep farming and agriculture, it may be assessed that shepherds and farmers are the working figures potentially at highest risk of exposure. The continuous movements of the flocks and the machining of the soil could cause suspension and diffusion of powders containing chrysotile, tremolite and actinolite fibers, with the consequent risk of inhalation by people employed in these activities. A similar situation could occur for farmers who, meanwhile carrying out their business, can breathe dangerous chrysotile, tremolite and actinolite fibers. Furthermore, other figures can suffer damage as a result of exposure to asbestos dust that is freed during the construction of rural buildings (for example houses) or other construction works (e.g., dirt roads and fences). Moreover, due to its geomorphological, geological and climatic setting, the Basilicata region (Figure 1) is affected by the diffuse presence of landslides [45] that may disturb NOA-bearing outcrops. Indeed, chrysotile together with asbestos tremolite and actinolite may release airborne dust in the neighboring environment, thus increasing population exposure to hazardous air fibers.

4. Conclusions

In this study, serpentinite rocks cropping out nearby the Episcopia village (Lucania, Southern Italy) have been investigated by means of various analytical techniques (i.e., XRPD, SEM-EDS and DTG) with the aim to assess the occurrence of naturally occurring asbestos. The results obtained indicate that the presence of asbestos was detected in all the serpentinite samples, and therefore it may be deduced that all of the analyzed specimens are potentially injurious to human health.

The presence of chrysotile was detected in nine of the eleven samples analyzed, while asbestos tremolite and asbestos actinolite were identified in five and three samples, respectively. The observed dimension of these fibers generally matched with the size of regulated asbestos. Weathering processes and/or human activities are able to produce dust containing asbestos fibers which are potentially inhalable, therefore increasing the human health risks. Their wide dispersion into the environment makes inhalation a risk even for those people not related to occupational purposes. The presence of NOA during working activities should be considered in the preliminary planning step to avoid workers' health risks and sanitary risks for the population living near asbestos sites. The disturbance (excavations, remediation, and moving) of asbestos potentially containing rocks and soils should be foreseen and planned so that adequate control measures may be carried out to avoid the spreading of airborne asbestos dust during work.

It is worth mentioning that, owing to possible health problems due to asbestos fiber dispersion the Italian law regulates these types of outcrops, demanding the asbestos presence identification in order to increase health safeguard. These new knowledge and highlights can be used to provide data for compulsory Italian mapping and should encourage local, regional and national authorities to avoid and to prevent asbestos exposure risks. Moreover, this study could be useful to make the population aware of the geological context in which they live, in order to take adequate prevention measures and good practices in everyday life. Therefore, the asbestos minerals investigation is essential from both scientific and legislative viewpoint, particularly for the administrative agencies, whose task it is to defense public health and to implement construction and safeguard policies.

Author Contributions: Conceptualization, A.B. and R.P.; methodology, A.B. and E.G.; software, A.B.; validation, A.B., R.P. and C.R.; formal analysis, A.B. and P.Z.; investigation, A.B., C.R. and R.P.; resources, A.B. and R.P.; data curation, A.B. C.R. and R.P.; writing—original draft preparation, A.B., C.R. and R.P.; writing—review and editing, A.B., C.R., R.P., P.Z., I.F., D.M., C.A., E.G., A.C. and R.D.R.; visualization, A.B.; supervision, A.B. and R.P.; project administration, A.B. and R.P.; and funding acquisition, A.B. and R.P.

Funding: The work has received financial support from the FFABR fund (by the Italian MIUR) scientific responsible Andrea Bloise. Part of this research was carried out under the financial support of "Piano Triennale della Ricerca (2017–2020)" (Università di Catania, Dipartimento di Scienze Biologiche, Geologiche e Ambientali), scientific responsible Rosalda Punturo.

Acknowledgments: The authors thank E. Barrese for the support during data collection. The work has received financial support from University of Calabria and University of Catania.

Conflicts of Interest: The authors declare no conflict of interest.

References

1. Baumann, F.; Buck, B.J.; Metcalf, R.V.; McLaurin, B.T.; Merkler, D.J.; Carbone, M. The presence of asbestos in the natural environment is likely related to mesothelioma in young individuals and women from Southern Nevada. *J. Thorac. Oncol.* **2015**, *10*, 731–737. [CrossRef] [PubMed]
2. Gamble, J.F.; Gibbs, G.W. An evaluation of the risks of lung cancer and mesothelioma from exposure to amphibole cleavage fragment. *Regul. Toxicol. Pharmacol.* **2008**, *52*, 154–186. [CrossRef]
3. Hillerdal, G. Mesothelioma: Cases associated with non-occupational and low dose exposures. *Int. J. Occup. Environ. Med.* **1999**, *56*, 505–513. [CrossRef]
4. Berry, G.; Gibbs, G.W. Mesothelioma and asbestos. *Regul. Toxicol. Pharmacol.* **2008**, *52*, S223–S231.
5. Gualtieri, A.F. *Mineral Fibers: Crystalchemistry, Chemical-Physical Properties, Biological Interaction and Toxicity*; European Mineralogical Union and Mineralogical Society of Great Britain and Ireland: London, UK, 2017; p. 533.
6. Punturo, R.; Bloise, A.; Critelli, T.; Catalano, M.; Fazio, E.; Apollaro, C. Environmental implications related to natural asbestos occurrences in the ophiolites of the Gimigliano-Mount Reventino Unit (Calabria, southern Italy). *Int. J. Environ. Res.* **2015**, *9*, 405–418.
7. Bloise, A.; Punturo, R.; Catalano, M.; Miriello, D.; Cirrincione, R. Naturally occurring asbestos (NOA) in rock and soil and relation with human activities: The monitoring example of selected sites in Calabria (southern Italy). *Ital. J. Geosci.* **2016**, *135*, 268–279. [CrossRef]
8. Vignaroli, G.; Ballirano, P.; Belardi, G.; Rossetti, F. Asbestos fibre identification vs. evaluation of asbestos hazard in ophiolitic rock mélanges, a case study from the Ligurian Alps (Italy). *Environ. Earth Sci.* **2014**, *72*, 3679–3698. [CrossRef]
9. Bloise, A.; Belluso, E.; Critelli, T.; Catalano, M.; Apollaro, C.; Miriello, D.; Barrese, E. Amphibole asbestos and other fibrous minerals in the meta-basalt of the Gimigliano-Mount Reventino Unit (Calabria, south-Italy). *Rend. Online Soc. Geol. Ital.* **2012**, *21*, 847–848.
10. Harper, M. 10th Anniversary critical review: Naturally occurring asbestos. *J. Environ. Monit.* **2008**, *10*, 1394–1408. [CrossRef] [PubMed]
11. Baumann, F.; Ambrosi, J.-P.; Carbone, M. Asbestos is not just asbestos: An unrecognised health hazard. *Lancet Oncol.* **2013**, *14*, 576–578. [CrossRef]
12. Ballirano, P.; Bloise, A.; Cremisini, C.; Nardi, E.; Montereali, M.R.; Pacella, A. Thermally induced behavior of the K-exchanged erionite: A further step in understanding the structural modifications of the erionite group upon heating. *Period. Mineral.* **2018**, *87*, 123–134.
13. Ballirano, P.; Pacella, A.; Bloise, A.; Giordani, M.; Mattioli, M. Thermal Stability of Woolly Erionite-K and Considerations about the Heat-Induced Behaviour of the Erionite Group. *Minerals* **2018**, *8*, 28. [CrossRef]
14. Cardile, V.; Lombardo, L.; Belluso, E.; Panico, A.; Capella, S.; Balazy, M. Toxicity and carcinogenicity mechanisms of fibrous antigorite. *Int. J. Environ. Res. Public Health* **2007**, *4*, 1–9. [CrossRef] [PubMed]
15. Pinizzotto, M.R.; Cantaro, C.; Caruso, M.; Chiarenza, L.; Petralia, C.; Turrisi, S.; Brancato, A. Environmental monitoring of airborne fluoro-edenite fibrous amphibole in Biancavilla (Sicily, Italy): A nine-years survey. *J. Mediterr. Earth Sci.* **2018**, *10*, 89–95. [CrossRef]
16. Bellomo, D.; Gargano, C.; Guercio, A.; Punturo, R.; Rimoldi, B. Workers' risks in asbestos contaminated natural sites. *J. Mediterr. Earth Sci.* **2018**, *10*, 97–106.
17. Ballirano, P.; Bloise, A.; Gualtieri, A.F.; Lezzerini, M.; Pacella, A.; Perchiazzi, N.; Dogan, M.; Dogan, A.U. The Crystal Structure of Mineral Fibers. In *Mineral Fibers: Crystal Chemistry, Chemical-Physical Properties, Biological Interaction and Toxicity*; Gualtieri, A.F., Ed.; European Mineralogical Union: London, UK, 2017; Volume 18, pp. 17–53.
18. Belluso, E.; Cavallo, A.; Halterman, D. Crystal habit of mineral fibres. In *Mineral Fibres: Crystal Chemistry, Chemical-Physical Properties, Biological Interaction and Toxicity*; Gualtieri, A.F., Ed.; European Mineralogical Union: London, UK, 2017; Volume 18, pp. 65–109.
19. Leake, B.E.; Woolley, A.R.; Arps, C.E.; Birch, W.D.; Gilbert, M.C.; Grice, J.D.; Linthout, K.; Laird, J.; Mandarino, J.; Maresch, W.V.; et al. Nomenclature of amphiboles: Report of the subcommittee on amphiboles of the international mineralogical association, commission on new minerals and mineral names. *Can. Mineral.* **1997**, *35*, 219–246.

20. Bloise, A.; Fornero, E.; Belluso, E.; Barrese, E.; Rinaudo, C. Synthesis and characterization of tremolite asbestos fibres. *Eur. J. Mineral.* **2008**, *20*, 1027–1033. [CrossRef]
21. Bloise, A.; Barca, D.; Gualtieri, A.F.; Pollastri, S.; Belluso, E. Trace elements in hazardous mineral fibres. *Environ. Pollut.* **2016**, *216*, 314–323. [CrossRef]
22. Bloise, A.; Catalano, M.; Critelli, T.; Apollaro, C.; Miriello, D. Naturally occurring asbestos: Potential for human exposure, San Severino Lucano (Basilicata, Southern Italy). *Environ. Earth Sci.* **2017**, *76*, 648. [CrossRef]
23. Punturo, R.; Ricchiuti, C.; Bloise, A. Assessment of Serpentine Group Minerals in Soils: A Case Study from the Village of San Severino Lucano (Basilicata, Southern Italy). *Fibers* **2019**, *7*, 18. [CrossRef]
24. Dichicco, M.C.; Paternoster, M.; Rizzo, G.; Sinisi, R. Mineralogical Asbestos Assessment in the Southern Apennines (Italy): A Review. *Fibers* **2019**, *7*, 24. [CrossRef]
25. Massaro, T.; Baldassarre, A.; Pinca, A.; Martina, G.L.; Fiore, S.; Lettino, A.; Cassano, F.; Musti, M. Exposure to asbestos in buildings in areas of Basilicata characterized by the presence of rocks containing tremolite. *G. Ital. Med. Lav. Ergon.* **2012**, *34* (Suppl. 3), 568–570.
26. Dichicco, M.C.; Laurita, S.; Paternoster, M.; Rizzo, G.; Sinisi, R.; Mongelli, G. Serpentinite Carbonation for CO_2 Sequestration in the Southern Apennines: Preliminary Study. *Energy Procedia* **2015**, *76*, 477–486. [CrossRef]
27. Caputo, A.; De Santis, M.; Manno, V.; Cauzillo, G.; Bruni, B.M.; Palumbo, L.; Conti, S.; Comba, P. Health impact of asbestos fibres naturally occurring in Mount Pollino area (Basilicata Region, Southern Italy). *Epidemiol. Prev.* **2018**, *42*, 142–150. [PubMed]
28. Pasetto, R.; Bruni, B.; Bruno, C.; Cauzillo, G.; Cavone, D.; Convertini, L.; De Mei, B.; Marconi, A.; Montagano, G.; Musti, M.; et al. Mesotelioma pleurico ed esposizione ambientale a fibre minerali: Il caso di un'area rurale in Basilicata. *Annali dell'Istituto Superiore di Sanità* **2004**, *40*, 251–265. [PubMed]
29. Musti, M.; Bruno, C.; Cassano, F.; Caputo, A.; Cauzillo, G.; Cavone, D.; Convertini, L.; De Blasio, A.; De Mei, B.; Marra, M.; et al. Sorve-glianza sanitaria delle popolazioni esposte a fibre di tremolite nel territorio della ASL 3—Lagonegro (PZ). *Annali dell'Istituto Superiore di Sanità* **2006**, *42*, 469–476. [PubMed]
30. Pieri, P.; Vitale, G.; Beneduce, P.; Doglioni, C.; Gallicchio, S.; Giano, S.I.; Loizzo, R.; Moretti, M.; Prosser, G.; Sabato, L.; et al. Tettonica quaternaria nell'area bradanico-ionica. *Il Quaternario* **1997**, *10*, 535–542.
31. Critelli, S.; Le Pera, E. Post-Oligocene sediment dispersal systems and unroofing history of the Calabrian Microplate, Italy. *Int. Geol. Rev.* **1998**, *48*, 609–637. [CrossRef]
32. Vitale, S.; Ciarcia, S.; Tramparulo, F.D.A. Deformation and stratigraphic evolution of the Ligurian Accretionary Complex in the southern Apennines (Italy). *J. Geodyn.* **2013**, *66*, 120–133. [CrossRef]
33. Apollaro, C.; Dotsika, E.; Marini, L.; Barca, D.; Bloise, A.; De Rosa, R.; Doveri, M.; Lelli, M.; Muto, F. Chemical and isotopic characterization of the thermo mineral water of Terme Sibarite springs (Northern Calabria, Italy). *Geochem. J.* **2012**, *46*, 117–129. [CrossRef]
34. Guagliardi, I.; Buttafuoco, G.; Apollaro, C.; Bloise, A.; De Rosa, R.; Cicchella, D. Using gamma-ray spectrometry and geostatistics for assessing geochemical behavior of radioactive elements in the Lese catchment (southern Italy). *Int. J. Environ. Res.* **2013**, *7*, 645–658.
35. Apollaro, C.; Fuoco, I.; Brozzo, G.; De Rosa, R. Release and fate of Cr (VI) in the ophiolitic aquifers of Italy: The role of Fe (III) as a potential oxidant of Cr (III) supported by reaction path modelling. *Sci. Total Environ.* **2019**, *660*, 1459–1471. [CrossRef]
36. Bonardi, G.; Amore, F.O.; Ciampo, G.; Capoa, P.; Miconnet, P.; Perrone, V. Il Complesso Liguride Auct.: Stato delle conoscenze e problemi aperti sulla sua evoluzione pre-appenninica ed i suoi rapporti con l'Arco Calabro. *Mem. Soc. Geol. Ital.* **1988**, *41*, 17–35.
37. Belviso, C.; Lettino, A.; Cavalcante, F.; Fiore, S.; Finizio, F. *Carta Geologica delle Unità Liguridi dell'area del Pollino (Basilicata)*; Digilabs: Bari, Italy, 2009; ISBN 978-88-7522-026-6.
38. Whitney, D.L.; Evans, B.W. Abbreviations for names of rock-forming minerals. *Am. Mineral.* **2010**, *95*, 185–187. [CrossRef]
39. Bloise, A.; Kusiorowski, R.; Lassinantti Gualtieri, M.; Gualtieri, A.F. Thermal behaviour of mineral fibers. In *Mineral Fibers: Crystal Chemistry, Chemical-Physical Properties, Biological Interaction and Toxicity*; Gualtieri, A.F., Ed.; European Mineralogical Union: London, UK, 2017; Volume 18, pp. 215–252.
40. Viti, C. Serpentine minerals discrimination by thermal analysis. *Am. Mineral.* **2010**, *95*, 631–638. [CrossRef]
41. Fubini, B.; Mollo, L. Role of iron in the reactivity of mineral fibers. *Toxicol. Lett.* **1995**, *82*, 951–960. [CrossRef]

42. Shukla, A.; Mossman, B. Cell signalling and transcription factor activation by asbestos in lung injury and disease. *Int. J. Biochem. Cell Biol.* **2003**, *35*, 1198–1209. [CrossRef]

43. Turci, F.; Tomatis, M.; Pacella, A. Surface and bulk properties of mineral fibres relevant to toxicity. In *Mineral Fibers: Crystal Chemistry, Chemical-Physical Properties, Biological Interaction and Toxicity*; Gualtier, A.F., Ed.; European Mineralogical Union: London, UK, 2017; Volume 18, pp. 171–214.

44. Oze, C.; Solt, K. Biodurability of chrysotile and tremolite asbestos in simulated lung and gastric fluid. *Am. Mineral.* **2010**, *95*, 825–831. [CrossRef]

45. Lazzari, M.; Piccarreta, M.; Capolongo, D. Landslide triggering and local rainfall thresholds in Bradanic Foredeep, Basilicata region (southern Italy). In *Landslide Science and Practice*; Springer: Berlin/Heidelberg, Germany, 2013; pp. 671–677.

![fibers logo] *fibers*

MDPI

Article

Characterization of Serpentines from Different Regions by Transmission Electron Microscopy, X-ray Diffraction, BET Specific Surface Area and Vibrational and Electronic Spectroscopy

Miguel A. Rivero Crespo [1], Dolores Pereira Gómez [2], María V. Villa García [3], José M. Gallardo Amores [4] and Vicente Sánchez Escribano [3,*]

[1] Laboratorium für Organische Chemie, ETH Zürich, Vladimir-Prelog-Weg 3, HCI, 809 Zürich, Switzerland; miguel_angel.rivero_crespo@org.chem.ethz.ch
[2] Departamento de Geología, Facultad de Ciencias, Universidad de Salamanca, Pl. de la Merced s/n, 37008 Salamanca, Spain; mdp@usal.es
[3] Departamento de Química Inorgánica, Facultad de Ciencias Químicas, Universidad de Salamanca, Pl. de la Merced s/n, 37008 Salamanca, Spain; mvilla@usal.es
[4] Laboratorio de altas presiones, Departamento de Química Inorgánica I, Universidad Complutense de Madrid, Avd. Complutense, s/n, 28040 Madrid, Spain; amores@quim.ucm.es
* Correspondence: vsescrib@usal.es

Received: 26 March 2019; Accepted: 14 May 2019; Published: 20 May 2019

Abstract: Serpentinite powdered samples from four different regions were characterized using scanning electron microscopy (SEM), X-ray diffraction (XRD), S_{BET} and porosity measurements, UV-Vis and Infrared Spectroscopy of the skeletal region and surface OH groups. SEM micrographs of the samples showed a prismatic morphology when the lizardite was the predominant phase, while if antigorite phase prevailed, the particles had a globular morphology. The few fibrous-shaped particles, only observed by SEM and weakly detected by XRD on MO-9C and MO13 samples, were characteristic of the chrysotile phase. All diffraction XRD patterns showed characteristic peaks of antigorite and lizardite serpentine phases, with crystallite sizes in the range 310–250 Å and with different degrees and types of carbonation processes, one derived from the transformation of the serpentine, generating dolomite, and another by direct precipitation of calcite. The S_{BET} reached values between 38–24 $m^2 \cdot g^{-1}$ for the samples less crystalline, in agreement with the XRD patterns, while those with a higher degree of crystallinity gave values close to 8–9 $m^2 \cdot g^{-1}$. In the UV region all electronic spectra were dominated by the absorption edge due to $O^{2-} \rightarrow Si^{4+}$ charge transfer transition, with Si^{4+} in tetrahedral coordination, corresponding to a band gap energy of ca 4.7 eV. In the visible region, 800–350 nm, the spectra of all samples, except Donai, presented at least two weak and broad absorptions centred in the range 650–800 and 550–360 nm, associated with the presence of Fe^{3+} ions from the oxidation of structural Fe^{2+} ions in the serpentinites $((Mg_xFe^{2+}_{1-x})_3Si_2O_5(OH)_4)$. The relative intensity of the IR bands corresponding to the stretching modes of the OH's groups indicated the prevalence of one of the two phases, antigorite or lizardite, in the serpentinites. We proposed that the different relative intensity of these bands could be considered as diagnostic to differentiate the predominance of these phases in serpentinites.

Keywords: XRD; SEM; IR; RD-UV-Vis spectroscopy; chrysotile; antigorite; lizardite characterization

1. Introduction

Serpentine minerals are phyllosilicates with a large metamorphic stability field that is generally formed during the hydration of basic to ultrabasic rocks. They have a general formula

$(Mg_xFe_{1-x})_3Si_2O_5(OH)_4$; $0 \le x \le 1$ and are characterized by a 1:1 unit with a tetrahedral Si-centred layer and a trioctahedral Mg-centred layer. The different spatial arrangements of these layers result in three main polymorphs: Lizardite, chrysotile and antigorite. The sheets form flats layers in lizardite, fibrous cylinders in chrysotile and corrugated structures in antigorite [1]. Although it is known that lizardite and chrysotile are the main varieties that are present in serpentinites with low-grade metamorphic transformations [2,3] and antigorite is found in rocks with high-grade metamorphic transformation environments [4], the P-T conditions of their thermodynamic transition are poorly defined and sometimes it is difficult to evaluate to which conditions the rocks were exposed. On the other hand, the variety of chemical substitutions of mono-, di and tri-valent metals ions [5,6] makes it difficult to specify the composition and thermal stability of these phases.

Serpentinites are commonly used as raw materials for the production of a high purity of magnesium oxide, asbestos and as ornamental stone [7]. In spite of this, details of their physicochemical properties such as the crystal structure of the most common phase lizardite and antigorite are still a matter of controversy. The behaviour of these rocks seems to be dependent on the serpentine phase that is predominant in their composition [8,9] although sometimes it is not easy to identify them by traditional petrographic methods. Additionally, the application of serpentinites depends upon the mineralogical composition [10] and a specific analysis of the mineralogy of these rocks could provide appropriate information about their possible use in industry.

In this paper, we reported the results of a characterization study by SEM microscopy, XRD and IR and Vis-UV spectroscopy of six different serpentines, one from India, one from Portugal and the other four serpentinites from two different regions in Spain that had been the subject of other investigations. Our interest in this study was to deepen the understanding of the physicochemical properties of this important family of materials and propose a quick and easy method in order to discern each of the phases that were mixed in these serpentine minerals.

2. Materials and Methods

Six natural serpentinites were collected from four different massifs, where they were quarried and marketed with the following commercial names:

Verde Pirineos (MO-9C and MO13) from Moeche, in Cabo Ortegal, Galicia (Spain) [8,9].

Verde Macael (MA8 and MA-2B) from Macael in Andalucía (Spain) [11,12].

Rain Forest Green (FG1), from Udaipur Rajasthan (India). This rock was extracted in a bedrock quarry near the village of Vidhasar in the district of Bikaner, Rajasthan, India. This serpentinite is very resistant to different strengths, compared to others of the same nature, such as Verde Donai (Donai), from the Bragança-Morais massif (Portugal).

The rock samples were grounded and prepared to get homogeneous polycrystalline samples in the case of X-ray analyses.

Thin sections of the six serpentinites were studied for a first mineralogical analysis using a Leica DM2500P microscope with digital camera.

Scanning electron microscopy (SEM) studies were recorded with a Zeiss EM 900 instrument. Previously, the samples were prepared in aqueous suspension in an ultrasonic bath.

X-ray powder diffraction (XRD) patterns of the six serpentines tested were recorded on a Siemens D-500 (40 kV and 30 mA) diffractometer using CuKa radiation (λ = 1.5437 Å) Ni-filter, equipped with the Diffract ATV3 software package. The crystal size was determined using the Scherrer formula [12].

The BET surface areas and porosity were measured using N_2 adsorption/desorption at 77 K determined both volumetrically with a Micromeritics Gemini 2380 instrument.

FT-IR studies were performed with a Nicolet Avatar 360 FT-IR spectrophotometer in the range 4000–400 cm^{-1}, recording each spectrum from the same amount of sample pressed into thin wafers with KBr and in an air atmosphere. Vis-UV spectra were recorded using a Shimadzu UV-2401 PC spectrophotometer in the range 200–800 nm using the diffuse reflectance technique (DR-UV-VIS).

3. Results and Discussion

3.1. Petrographic Studies

From the petrographic studies, we observed that all samples presented a high percentage of serpentine phases (Figure 1a,b). Some of them were extremely carbonated. The origin of carbonation in each case is being investigated at present.

Figure 1. (**a**) Serpentinite from Macael: Most original minerals have been replaced by serpentine. The latter was replaced by carbonates in a later stage and acicular crystals of amphibole were cutting the serpentine in a final metamorphic transformation. Cross Nichols microphotograph. (**b**) Serpentinite from Cabo Ortegal (Moeche). The original ultramafic rock completely transformed into serpentine. Fibrous chrysotile could be identified by cutting the other serpentine phases of the rock. Cross Nichols microphotograph.

3.2. Scanning Electron Micrographs (SEM)

SEM of the serpentine samples are reported in Figure 2. The morphology of the MA-2B and MA8 samples was mainly prismatic, corresponding to the lizardite phase, while the other samples, Donai, FG1, MO-9C, MO13, showed quite a globular morphology, in agreement with the literature data [1,4], probably due to the high-grade of metamorphic transformation of these samples with respect to the first two ones. In addition, the few particles with fibrous shapes observed in MO-9C and MO13 samples were assigned to the chrysotile phase of the serpentines [6].

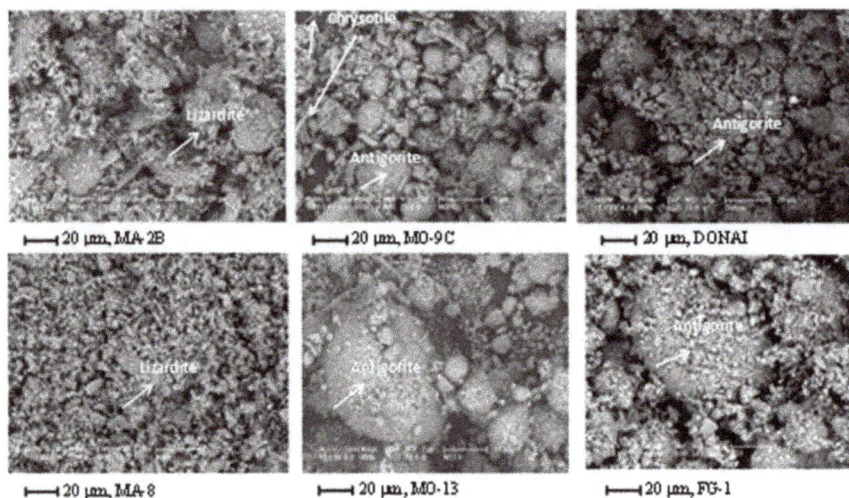

Figure 2. Scanning electron microscopy micrographs of serpentinites.

3.3. X-ray Diffraction

The diffraction patterns of the serpentine samples are compared in Figure 3 and their crystallographic parameters calculated by the Chekcell v.2. program of refinement software [13] is summarized in Table 1. All graphics showed characteristic peaks of serpentine-based structures with crystallite sizes calculated by the Scherrer method [14] in the range 310–250 Å. In MA8 and MA-2B samples the fundamental phase of the serpentines was lizardite ($Z = 1$, ICDD No. 11-0386) and their crystallographic parameters agreed with those reported in the ICDD card. The pattern for the MA8 sample corresponded mainly to the lizardite phase [12–15], with principal (h k l) reflections at 2θ values: 12.2 (0 0 1), 24.6 (0 0 2), 35.5 (1 1 1), 41.9 (1 0 3) and 61.5(1 1 1) while in MA-2B samples additional reflections were observed and its diffractions peaks were broader due to its lower crystallinity and crystallite size. In the case of the MA-2B sample the new peaks were assigned to $Ca(OH)_2$ (ICDD No.78-0315, $2\theta = 18.8$ and 34.7), calcite ($CaCO_3$), (ICDD No. 5-586, $2\theta = 29.4, 39.6, 44.8, 47,6$ and 48.3) and dolomite ($Ca_{1-x}Mg_x(CO_3)$), (ICDD No. 36-426, $2\theta = 30.8$). Taking into account the XRD analysis in these samples, it seemed that there were two types of carbonation processes, one derived from the transformation of the serpentine, generating dolomite, and another from direct precipitation of calcite. It should be noted that in the area of origin of Macael serpentinites some of them appeared sandwiched between sheets of carbonates. Ca^{2+} ions coming from these layers, combined with CO_2 fluids circulating on fractures and produced calcite precipitates [11].

In samples MO-9C and MO13 from Cabo Ortegal (Galicia) as well as Donai from Bragança (Portugal) and FG1 from India, the predominant polymorph of serpentine was antigorite [16] ($Z = 6$, ICDD No. 10-0402); i.e., serpentine from the transformation of surface olivines and pyroxenes (probably due to the influence of meteoric water and mainly of sea water) whose crystallographic parameters agreed with those reported in the ICDD card. Four samples had a content higher than 75% of antigorite, but carbonation was greater in FG1 and MO13 and, as a result, besides the peaks corresponding to the antigorite, other small diffraction lines could be observed.

Patterns for MO-9C and Donai samples corresponded mostly to antigorite serpentine mineral with principal (h k l) reflections at 2θ values: 12.1 (0 0 6), 24.3 (0 0 12), 35.4 (1 3 4) and 60.1 (2 3 18). In the case of FG1, MO13 and MO-9C, in the latter with less intensity, additional reflections from dolomite (ICDD No. 36-426, 2θ = 30.8), CaO (ICDD No. 82-1691, 2θ = 32.3) and MgO (ICDD No. 78-0430, 2θ = 42.9) were observed.

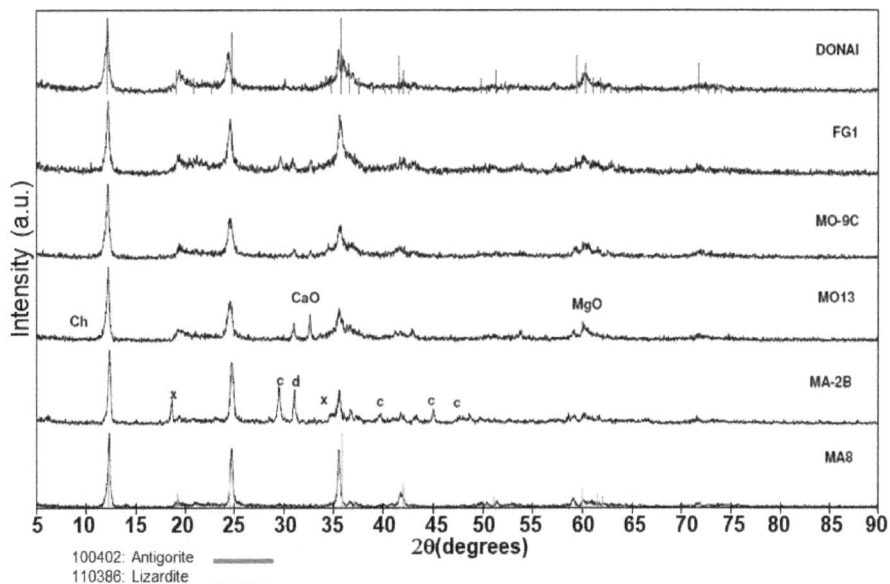

Figure 3. XRD patterns of serpentinites: "ch" chrysotile, "x" Ca(OH)$_2$, "c" calcite and "d" dolomite. The position and intensities of the peaks were reported according to the ICDD cards.

Table 1. Mineralogic composition, crystallographic data, calculated from the XRD patterns and specific surface area.

Sample	Mineral Phase	Cell Paramaters(Å)			V(Å^3)	D(Å)	S$_{BET}$ (m^2·g^{-1})
		A(Å)	B(Å)	C(Å)			
MA8	*Lizardite*, calcite, dolomite, magnetite	5.335(3)	5.335(3)	7.335(3)	180.8(3)	308	8
MA-2B	*Lizardite*, calcite, dolomite, clorite, tremolite	5.329(3)	5.329(3)	7.339(2)	180.5(2)	257	9
MO-9C	*Antigorite*, magnesite, dolomite	5.2999(4)	9.217(1)	42.910(4)	2096.2(6)	280	32
MO13	*Antigorite*, chrysotile, magnesite, dolomite	5.2892(6)	9.204(2)	42.862(6)	2086.7(9)	307	25
FG1	*Antigorite*, calcite, dolomite	5.2919(6)	9.214(2)	42.786(5)	2086.3(9)	380	25
Donai	*Antigorite*	5.301(1)	9.2064(2)	42.726(6)	2085.1(9)	237	38

Lizardite ICDD file No. 18-0779: a = 5.73172; monoclinic, SG: P31m (157), Z = 1. *Antigorite* ICDD file No. 12-0583: a = 5.305, b = 9.189, c = 42.75, orthorombic, SG: (O), Z = 6. Predominant phase is marked in *italics*.

3.4. Surface Area and Porosity

The surface area values for all serpentine samples are reported in Table 1. The nitrogen adsorption/desorption isotherms (Figure 4) corresponded to type II of the BDDT classification [17]. Desorption and adsorption branches were coincident, which showed the non-existence of mesopores but the presence of macropores (diameter, ø > 50 nm) in these materials [17].

The specific surface areas (S_{BET}) reached values between 38–24 $m^2 \cdot g^{-1}$ for the samples less crystalline (Donai, F1G and MO-9C and Mo-13), in agreement with XRD patterns, while those with values of 8–9 $m^2 \cdot g^{-1}$ corresponded to the higher degree of crystallinity (MA8 and MA-2B) in accordance with its lower surface area.

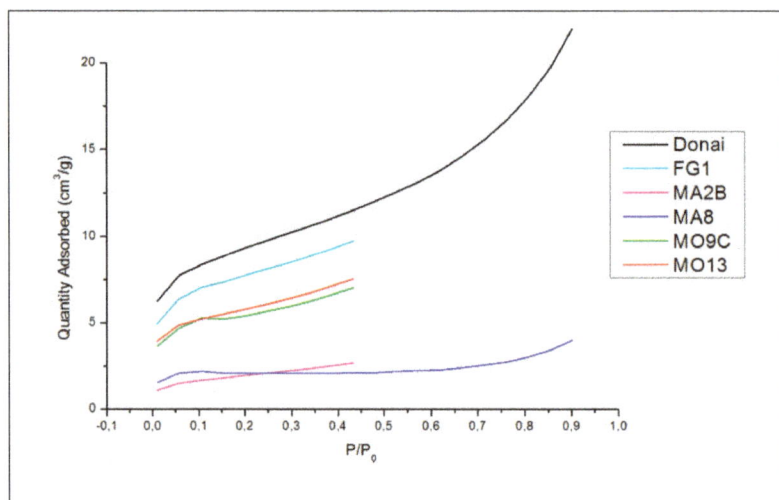

Figure 4. Nitrogen adsorption isotherms on serpentinite samples at 77 K.

3.5. DR-UV-Vis Spectroscopy

The electronic spectra of the investigated samples are shown in Figure 5. All spectra were dominated by the absorption edge associated with O^{2-} (2p) → Si^{4+} (3p) charge transfer transition, with Si^{4+} in tetrahedral coordination, in the range 350–265 nm, corresponding to a band gap energy of ca 4.7 eV. The E_g energy gap of MgO corresponded to charge transfer O^{2-} → Mg^{2+} was reported [18] to be of ca. 8 eV, corresponding to wavelength of 155 nm, which was just below the lower wavelength limit of our instrument.

In the visible region, 800–350 nm, all spectra, except Donai, presented at least two weak and broad absorptions centred in the regions 650–800 and 550–360 nm. According to previous studies [19–21] the first band could be associated with the $^6A_1$ → $^4T_1$ (^4G) crystal-field d → d transition of octahedral Fe^{3+} ions, while the second was due to an oxygen-to-metal charge-transfer transition (O^{2-} → Fe^{3+}). The absence of absorption bands in the visible region in the Donai sample indicated that it either lacked Fe^{2+} replacing Mg^{2+} ions or it had not been oxidized to Fe^{3+}.

The UV-VIS spectrum of the FG1 sample presented the same absorptions described above, although in this case the bands due to O^{2-} → Si^{4+} charge-transfer transition and d-d transition of octahedral Fe^{3+} ions, showed lower intensity than in the other ones. However, the absorption in the 500–350 nm range, where the intensity was higher than in the other samples, had two of the most intense components with maxima at 500 and 370 nm attributed to a charge-transfer transition (O^{2-} → Fe^{3+}) and metal-to-metal charge-transfer transition, which could be schematized as $2Fe^{3+}$ → Fe^{2+} + Fe^{4+}, respectively [22,23]. These data, according to XRD patterns, indicated a higher structural degradation of the FG1 sample with respect to the other ones and, as a result, an increase in the concentration of Fe^{3+} ions, due to oxidation of structural Fe^{2+} ions $(Mg_x,Fe^{2+}_{1-x})_3Si_2O_5(OH)_4)$ to Fe^{3+}, and a decrease in the concentration of Si^{4+} ions tetrahedrally coordinated in the serpentinite.

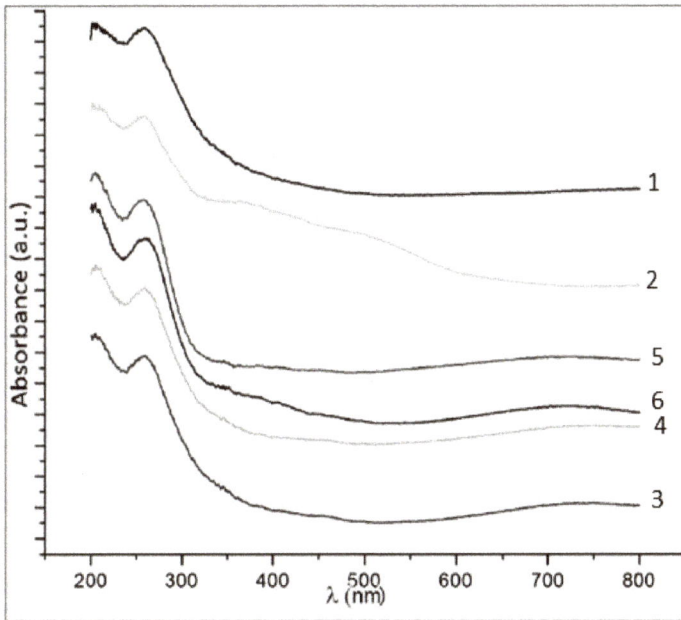

Figure 5. DR-UV–vis spectra of the different serpentinites: (1) DONAI, (2) FG1, (3) MO-13, (4) MO-9C, (5) MA-2B and (6) MA8.

3.6. FT-IR Spectroscopy

The IR spectra of the serpentine samples, in the 4000–400 cm^{-1} range, were characterized by three groups of bands: (i) Strong absorption bands in the range 3800–3500 cm^{-1} (Figure 6), due to MO–H (M = Fe^{2+} or Mg^{2+}) stretching vibrations modes; (ii) a very strong complex of bands in the range 1200–600 cm^{-1} (Figure 7), due to asymmetric and symmetric Si–O–Si stretching modes and (iii) one or more strong bands in the region 600–400 cm^{-1} (Figure 7) assigned to M–OH stretching (M = Fe^{2+} or Mg^{2+}) and Si-O-Si deformation modes [24]. Moreover, all the samples, except the so-called Donai, presented absorption bands characteristic of carbonate species [25] of varying intensity, near 1440 cm^{-1} and 880 cm^{-1}, due to symmetric stretching and bending modes, respectively (Figure 7).

The IR spectra of the six serpentinites in the region of the surface hydroxyl is recorded in Figure 6. All spectra were dominated by a broad absorption band with a maximum and shoulders of different relative intensities. In the case of Donai, MO-9C, FG1 and MO13, maxima at 3685 (for Donai) and 3681 cm^{-1} and shoulders at 3704, 3693 and 3670 cm^{-1} were observed. These bands are characteristic of the stretching modes of the isolated surface OH's over MgO_6 octahedra [26,27]. The spectra of the MA-2B and MA8 samples showed the same bands but the relative intensities differed significantly from the preceding ones. In particular, the spectra clearly showed a maximum at 3670 cm^{-1} and evident shoulders at 3704, 3693 and 3685 cm^{-1} i.e., the maximum at 3685 cm^{-1} and the shoulder at 3670 cm^{-1} in the former samples and in the two latter ones were observed as a shoulder and a maximum, respectively. The IR spectra corresponding to OH's stretching modes we reported here for serpentinite samples agreed with those reported in the literature [25,26]. However, in the case of our serpentinites, we observed clearly a shift in the maximum of absorption to lower wavenumbers from 3685–3681 cm^{-1} for Donai, FG1, MO13 and MO9C samples, in which the prevalent phase was antigorite, down to 3670 cm^{-1} for MA-8 y Ma-2B samples, where the prevalent phase was lizardite. We considered that this shift of the maximum of absorption, corresponded to the isolated surface OH's groups over MgO$_6$ octahedra, which was related to the structural effect of the antigorite or lizardite phases. In fact, the lower values of cell parameters in lizardite compared to antigorite corresponded to a lower Mg–O bond length, therefore in lizardite O–H bond length increased and as a result, the OH's stretching modes of the two rich in lizardite samples, MA-8 and MA-2B, were observed at a lower frequency than in the case of the other ones in which the predominant phase was antigorite.

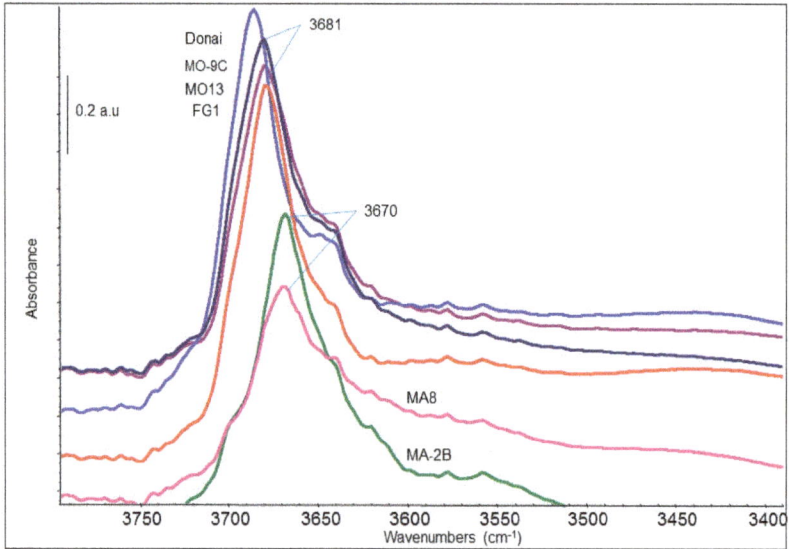

Figure 6. FT-IR spectra of the serpentinites in the 3800–3350 cm^{-1} range.

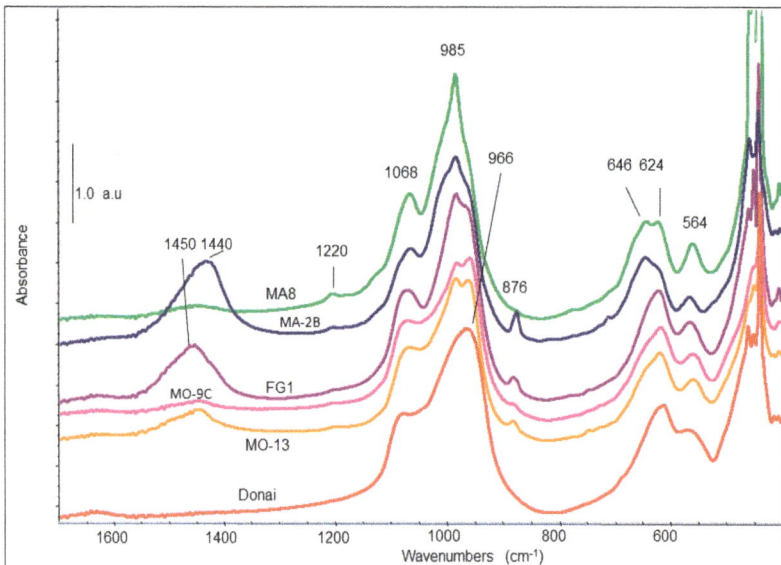

Figure 7. FT-IR spectra of the serpentinites in the 1700–400 cm^{-1} range.

The IR spectra in the skeletal region of all samples reported in Figure 7 could be interpreted on the basis of the known IR spectra of the silica polymorphs, where any oxygen atom in a tetrahedral-based SiO_2 polymorph was bonded with two tetrahedral cations (Si^{4+}) and has a point group of symmetry C_{2V} [28–30].

The spectra of MA8, MA-2B samples were composed of sharp bands, indicating a high ion ordering in the structure. These spectra showed the very strong Si–O–Si asymmetric stretching mode with a maximum at 985 cm^{-1} and shoulders at 1068 cm^{-1} (pronounced) and 966 cm^{-1}. The higher vibration mode, according to the literature [24,31], was due to the in-phase coupling of the asymmetric stretching modes of the nearest Si–O–Si groups. The spectra of the samples FG1, MO13, MO-9C, and Donai (Figure 7) were similar to the samples MA8 and MA-2B, although with less intense bands and different relative intensities. Thus, the sharp band observed on MA8 and MA-2B at 985 cm^{-1}, associated to asymmetric stretching of the Si–O–Si bridges decreased consecutively in intensity in FG1, MO13, MO-9C and Donai samples and, in parallel, the shoulder observed on the MA8 and MA-2B samples to 966 cm^{-1} increased its intensity up to be observed as an absolute maximum in the latter four samples. Considering that the absorption maxima corresponding to the O–Si–O asymmetric stretching mode were related to the value of cell parameters of the lizardite and antigorite phases, we associated the higher vibration frequency of the MA8 and MA-2B samples, i.e, lower Si–O–Si bond length, to the predominance of lizardite over antigorite and, on the contrary, the lower vibration frequency observed for the other four samples where antigorite predominated over lizardite, was in agreement with the SEM and DRX analysis.

In the 700–500 cm^{-1} region, two bands were observed, the first one, of medium strength with two components at 646 and 624 cm^{-1}, corresponded to the symmetric Si–O–Si stretching mode, which also had a Si–O–Si in-plane bending character, while the other one was observed at ca. 564 cm^{-1} was assigned to stretching Mg–OH$_{(vM-O)}$ in octahedral sites [24].

The first component at 646 cm^{-1} appeared as a maximum in samples MA8 and MA-2B, where the predominant phase was lizardite, while the second at 624 cm^{-1} was a maximum when the prevalent phase was antigorite. The maximum at 564 cm^{-1} observed for the MA8 and MA-2B samples shifted down (up to 555 cm^{-1}) for the F1G, MO13, MO-9C and Donai samples due to the higher Mg–OH bond energy of the two former samples with respect to the latter ones.

Considering a tetrahedral symmetry for SiO_4, the two components observed on the lowest frequency sharp band at 459 and 445 cm^{-1} were associated to the Si–O–Si out-of-plane bending vibration ($\delta_{Si-O-Si}$) and rocking mode of the Si–O–Si bridges [31], respectively.

The broad absorption band with maxima between 1430 and 1450 cm^{-1} and the sharp band at 876 cm^{-1} corresponding to the carbonate groups, decreased in intensity in the order: MA-2B, FG1 and MO13 samples, agreeing with the diffractograms of these samples, where, in addition to the antigorite and lizardite phases, diffraction peaks corresponding to the calcite and dolomite carbonates were also observed.

Particularly, in the MA8 sample the presence of two bands at 1445 and at ca. 1220 cm^{-1} (weak) characteristic of surface hydrogen carbonates [32] (C–O symmetric stretching and COH deformation modes, respectively), confirmed the formation of these species, more soluble than the carbonates, by reaction of carbonate species with excess CO_2 circulating through the fractures of the rocks, according to the reaction:

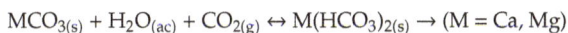

$$MCO_{3(s)} + H_2O_{(ac)} + CO_{2(g)} \leftrightarrow M(HCO_3)_{2(s)} \rightarrow (M = Ca, Mg)$$

4. Implications of the Serpentinites

Many different tools have been used in the study of serpentinites and their mineralogy. Most of them are confusing when trying to differentiate the different polymorphs. However, the proper mineralogical characterization of these rocks is important when used in different applications, from materials for construction to materials for insulation and as ornamental stone. Some fibrous minerals have been proven to be related to specific health problems such as asbestosis. It has been reported that

Fibers **2019**, *7*, 47

rigid fibers such as those identified with acicular amphiboles (e.g., actinolite, tremolite) can cause tissue damage that can degenerate into mesothelioma [33,34]. It has not been fully proven that serpentine polymorphs are related to health issues, but it is very important to present a complete and reliable characterization of stones when they are commercialised in a quality demanding market. The easy methods we proposed in this paper for the analysis of serpentinites such as Vibrational and Electronic spectroscopy can save a lot of time in the characterization of these rocks, with the implications in the advance of petrogenetic research of them, but also the economic implications related to different economic areas, from natural stone exploitation to the use of serpentine phases in new ceramic products, avoiding acicular/fibrous and suspicious phases.

The presence of Fe^{3+} cations, from the transformation of serpentinites, may be detected by Vis-UV spectroscopy. Thus, the presence of two broad absorptions bands at 800 and 350 cm, due to $^6A_1 \rightarrow {}^4T_1$ (^4G) crystal-field d $\rightarrow$ d transition of octahedral Fe^{3+} ions, and oxygen-to-metal charge-transfer transition ($O^{2-} \rightarrow Fe^{3+}$) was indicative of the presence of Fe^{3+} ions from the oxidation of octahedral Fe^{2+} ions, partially replacing Mg^{2+} ions, in the serpentine phase.

We proposed that the two bands of the IR spectrum assigned to terminals OH's groups on octahedral Mg^{2+} ions of serpentinites, with maxima at ca. 3685–3681 cm^{-1} for samples whose predominant phase was antigorite and 3670 cm^{-1} for those whose predominant phase was lizardite, could be considered diagnostic in order to differentiate the predominance of these phases in serpentinites.

Author Contributions: All authors have been involved in the analytical work, the writing and editing of the manuscript.

Funding: This research was partially funded by University of Salamanca, grant Ref: ref. KAW9/463AC01.

Acknowledgments: The authors thank the University of Salamanca under Grant ref. KAW9/463AC01 for partial financial support as well as the Electronic Microscopy Service of this university. Authors acknowledge the comments and advice of two anonymous reviewers that helped to enrich the original paper.

Conflicts of Interest: The authors declare no conflicts of interest.

References

1. Wicks, F.J.; O'Hanley, D.S. Serpentine Mineral: Structure and petrology. Reviews in Mineralogy. In *Hydrous Phylosilicates*; Baley, S.W., Ed.; Mineralogical Society of America: Chantilly, VA, USA, 1988; pp. 91–167.
2. Andreani, M.; Mevel, C.; Boulier, A.M.; Escertín, J. Dynamic control on serpentine crystallisation in veins: Contrainints on hydration process in oceanic peridotites. *Geochem. Geophys. Geosyst.* **2007**, *8*. [CrossRef]
3. Dicchico, M.C.; Paternoster, M.; Rizzo, G.; Sinisi, R. Mineralogical Asbesto Assessment in the Southern Appenines (Italy): A review. *Fibers* **2019**, *7*, 24. [CrossRef]
4. Auzende, A.L.; Devouard, B.S.; Danile, I.; Baronnet, A.; Lardeaux, J.M. Serpentines from Central Cuba: Petrology and HRSEM study. *Eur. J. Mineral.* **2002**, *14*, 905–914. [CrossRef]
5. Evans, B.W. The serpentinite multisysSEM revisited: Chrysotile is metastable. *Int. Geol. Rev.* **2004**, *46*, 479–506. [CrossRef]
6. Cressey, B.A.; Whittaker, E.J.W. Five-fold symmetry in chrysotile asbestos revealed by transmission electron microscopy. *Mineral. Mag.* **1993**, *5*, 729–732. [CrossRef]
7. Pereira, D.; Yenes, M.; Blanco, J.A.; Peinado, M. Characterization of serpentinites to define their appropriate use as dimension stone. *Geol. Soc. Lond. Spec. Publ.* **2007**, *271*, 55–62. [CrossRef]
8. Pereira, D.; Peinado, M.; Blanco, J.A.; Yenes, M. Geochemical characterization of a serpentinization process at Cabo Ortegal (NW Spain). *Can. Mineral.* **2008**, *46*, 317–327. [CrossRef]
9. Pereira, D.; Peinado, M.; Yenes, M.; Monterrubio, S.; Nespereira, J.; Blanco, J.A. Serpentinites from Cabo Ortegal (Galicia, Spain): A search for correct use as ornamental stones. *Geol. Soc. Lond. Spec. Publ.* **2010**, *333*, 81–85. [CrossRef]
10. Post, J.L.; Borer, L. High-resolution infrared spectra, physical properties, and micromorphology of serpentines. *Appl. Clay Sci.* **2000**, *16*, 73–85. [CrossRef]
11. Navarro, R.; Pereira, D.; Gimeno, A.; del Barrio, S. Verde Macael: A serpentinite wrongly referred to as a marble. *Geosciences* **2013**, *3*, 102–113. [CrossRef]

12. Navarro, R.; Pereira, D.; Rodríguez-Navarro, C.; Sebastián Pardo, E. The Sierra Nevada Serpentinites: The serpentinites most used in Spanish Heritage buildings. *Geol. Soc. Lond. Spec. Publ.* **2015**, *407*, 38–44. [CrossRef]
13. CHEKCELL.v.2. Program of Refinement Software by Jean Lauguier et Bernard Bochu. Available online: http://www.inpg.fr/LMGPorHttp://www.ccp14.ac.uk/tutorial/lmgp/ (accessed on 29 April 2019).
14. West, A.R. *Solid State Chemistry and Its Applications*; Wiley: Chichester, UK, 1996; p. 174.
15. Liu, K.; Chen, Q.; Hu, H.; Yin, Z. Characterization and leaching behaviour of lizardite in Yuanjiang laterite ore. *Appl. Clay Sci.* **2010**, *47*, 311–316. [CrossRef]
16. Li, W.; Huang, Z.; Liu, Y.; Fang, M.; Ouyang, X.; Huang, S. Phase behavior of serpentine mineral by carbothermal reduction nitridation. *Appl. Clay Sci.* **2012**, *57*, 86–90. [CrossRef]
17. Gregg, S.J.; Sing, K.S.W. *Adsorption Surface Area and Porosity*; Academic Press: London, UK, 1991.
18. Barry, C.; Norton, M.G. *Ceramic Materials: Science and Engineering*; Springer: New York, NY, USA, 2007.
19. Haar De, L.G.J.; Blase, G. Photoelectrochemical properties of MgTiO3 and other titanates with the ilmenite structure. *J. Electrochem. Soc.* **1985**, *132*, 2933. [CrossRef]
20. Willey, R.J.; Oliveri, S.A.; Busca, G. Structure and magnetic properties of magnesium ferrite fine powdersStructure and magnetic properties of magnesium ferrite fine powders. *J. Mater. Res.* **1993**, *8*, 1418–1427. [CrossRef]
21. Prieto García, M.C.; Gallardo Amores, J.M.; Sánchez Escribano, V.; Busca, G. Characterization of coprecipitated Fe_2O_3-Al_2O_3 powders. *J. Mater. Chem.* **1994**, *4*, 1123–1130. [CrossRef]
22. Kennedy, J.H.; Frese, K.W. Photooxidation of water at α-Fe_2O_3 electrodes. *J. Electrochem. Soc.* **1978**, *125*, 709. [CrossRef]
23. Busca, G.; Ramis, G.; Prieto García, M.C.; Sánchez Escribano, V. Preparation and characterization of $Fe_{2-x}Cr_xO_3$ mixed oxide powders. *J. Mater. Chem.* **1993**, *3*, 665–673. [CrossRef]
24. Mazza, D.; Lucco-Borlera, M.; Busca, G.; Delmastro, A. High-quartz solid-solution phases from xerogels with composition $2MgO.2Al_2O_3.5SiO_2$ μ-cordierite. *J. Eur. Ceram. Soc.* **1993**, *11*, 299–308. [CrossRef]
25. Bellamy, L.J. *The Infrared Spectra of Complex Molecules*, 3rd ed.; John Wiley and Sons, Inc.: New York, NY, USA, 1975.
26. Foresti, E.; Formaro, E.; Lesci, I.G.; Zuccheri, R.T.; Roveri, N. Asbestos health hazard: A spectroscopic study of synthetic geoinspired Fe-doped chrysotile. *J. Hazard. Mater.* **2009**, *167*, 1070–1079. [CrossRef]
27. Frost Ray, L.; Jagannadha Reddy, B.; Bahfenne, S.; Graham, J. Mid-infrared and near-infrared spectroscopic study of selected magnesium carbonate minerals containing ferric iron-Implications for the geosequestration of greenhouse gases. *Spectrochem. Acta Part A Mol. Biomol. Spectrosc.* **2009**, *72*, 597–604. [CrossRef]
28. Gadsden, J.A. *Infrared Spectra of Minerals and Related Inorganic Compounds*; Butterworths: London, UK, 1975.
29. Griffith, W.P. *Spectroscopy of Inorganic-Based Materials*; Clark, R.J.H., Hester, R.E., Eds.; Wiley: Hoboken, NZ, USA, 1987.
30. Astorino, E.; Peri, J.B.; Willey, R.J.; Busca, G. Spectroscopic characterization of silicate-1 and titanium silicate-1. *J. Catal.* **1995**, *157*, 482–500. [CrossRef]
31. Kamitsos, E.I.; Patsis, A.P.; Kordas, G. Infrared-reflectance spectra of heat-treated sol-gel-derived silica. *Phys. Rev. B* **1993**, *48*, 12499. [CrossRef]
32. Montanari, T.; Gastaldi, L.; lietti, L.; Busca, G. Basic catalysis and catalysis assisted by basicity: FT-IR and TPD characterization of potassium-doped alumina. *Appl. Catal. A Gen.* **2011**, *400*, 61–69. [CrossRef]
33. Dicchico, M.C.; de Bonis, A.; Mongelli, G.; Rizzo, G.; Sinisi, R. μ-Raman spectroscopy and X-ray diffraction of asbestos minerals for geoenvironmental monitoring: The case of the southern Appenines natural sources. *Appl. Clay Sci.* **2017**, *141*, 292–299. [CrossRef]
34. Dicchico, M.C.; Laurita, S.; Sinisi, R.; Battiloro, R.; Rizzo, G. Environmental and Health: The importance of tremolite occurence in the Pollino Geopark (Southern Italy). *Geosciences* **2018**, *8*, 98. [CrossRef]

fibers

MDPI

Article

Grinding Test on Tremolite with Fibrous and Prismatic Habit

Oliviero Baietto *, Mariangela Diano, Giovanna Zanetti and Paola Marini

Department of Environment, Land and Infrastructure Engineering (DIATI), Politecnico di Torino, 10129 Turin, Italy; mariangela.diano92@gmail.com (M.D.); giovanna.zanetti@polito.it (G.Z.); paola.marini@polito.it (P.M.)
* Correspondence: oliviero.baietto@polito.it

Received: 16 April 2019; Accepted: 20 May 2019; Published: 1 June 2019

Abstract: The main objective of this work is the evaluation of the morphology change in tremolite particles before and after a grinding process. The crushing action simulates anthropic alteration of the rock, such as excavation in rocks containing tremolite during a tunneling operation. The crystallization habit of these amphibolic minerals can exert hazardous effects on humans. The investigated amphibolic minerals are four tremolite samples, from the Piedmont and Aosta Valley regions, with different crystallization habits. The habits can be described as asbestiform (fibrous) for longer and thinner fibers and non-asbestiform (prismatic) for prismatic fragments, also known as "cleavage" fragments. In order to identify the morphological variation before and after the grinding, both a phase contrast optical microscope (PCOM) and a scanning electron microscope (SEM) were used. The identification procedure for fibrous and prismatic elements is related to a dimensional parameter (length–diameter ratio) defined by the Health and Safety Executive. The results highlight how mineral comminution leads to a rise of prismatic fragments and, therefore, to a potentially safer situation for worker and inhabitants.

Keywords: tremolite; Naturally Occurring Asbestos (NOA); asbestos; grinding test; PCOM; image analysis

1. Introduction

The object of this study is tremolite, a hydrated calcium magnesium silicate ($Ca_2Mg_5Si_8O_{22}(OH)_2$), belonging to the tremolite-ferro-actinolite series [1,2], with an amphibolic structure characterized by long and parallel double chains of silica tetrahedral (SiO_4) with a strip of cations located between the double chains [2–6].

Tremolite and actinolite can crystallize with two different crystalline habits and it is common to find them in fibrous habit, known as asbestiform, or in prismatic habit, non-asbestiform. This distinction can be described as [7]:

- "Asbestiform habit" is associated with a crystalline structure characterized by thin crystals similar to the morphology of organic fibers (hair; the resemblance is not in the width, 10^{-6} m for asbestos fibers vs 10^{-5} m for hair) or as a crystalline aggregation consisting of parallel fibers (bundles with indented extremities). The fibers are thin, long, and similar to needle-shaped elements with a unidirectional growth [8].
- "Non-asbestiform" refers to a structure characterized by tiny or "elongated prisms with a lozenge-shaped cross section" [4]. The crystalline growth is not unidirectional.

A crushing event acting on tremolite crystals could have different effects on crystalline habits, as illustrated by Illgren et al. [5]. In fact, amphiboles with asbestiform habits have a significant propensity to be longitudinally split. The longitudinal separation produces "fibrils" that are thinner and thinner without any perpendicular breakup to the elongation; therefore, fibers maintain their flexibility and tensile strength [4,5,9]. Non-asbestiform habits, however, consist of amphibolic minerals with internal cleavage. This term relates to the fact that breakage occurs along preferential planes, especially planes of relative weakness, and mainly perpendicular to the length [5,7]. The prismatic elements have, therefore, a tendency to fracture along these preferential planes, producing stocky prisms or acicular fragments with a reduction in their flexibility.

The particle morphology, size, physical–chemical properties and biopersistence correlated to the crystalline habits can involve relevant consequences to the human respiratory system. The Italian Minister of Health [10] highlighted the difficult removal process for asbestiform amphiboles, caused by the length and persistence of the fiber. These fibers are stronger and more flexible than cleavage fragments, so they tend to bend without breaking and also have negative repercussions on the defense mechanism operated by macrophages. Otherwise, Illgren et al. [5] specified that non-asbestiform (or prismatic) amphiboles have weaknesses and fragile behavior. Therefore, the prismatic component can be reduced into fragments that can progressively be cleared from the body by the macrophage phagocytosis.

Asbestos is included in Group 1, referring to "substances carcinogenic to humans", by the International Agency for Research on Cancer (IARC). It is important that the concept of regulated asbestos fibers, which correspond to fibers defined as "respirable" by the World Health Organization (WHO) (fibers having length > 5 μm, diameter ≤ 3 μm, and length/diameter (aspect) ratio ≥ 3:1) [11–13]. For this reason, many trials were focused on the evaluation of the carcinogenic effect, such as the potential for mesothelioma induction, for a non-asbestiform tremolite by means of in vivo (animals or human) and in vitro tests. These studies, seen in Table 1, demonstrate a negative or lower rate of carcinogenicity for non-asbestiform tremolite compared to asbestiform tremolite.

Table 1. Carcinogenic experiment on asbestiform and non-asbestiform tremolite.

Test Subjects	Procedure		Authors
Workers	Exposure to non-asbestiform tremolite		Gamble et al. [2]
In vitro	Test by a variety of cellular endpoints		Timbrell et al. [14] Wylie & Mossman [15] Wagner et al. [16]
In vivo (such as rat)	Experiment:	Inhalation	Davis et al. [17]
		Intraperitoneal or intrapleural injection	Wagner et al. [18] Smith et al. [19] Davis et al. [20]
		Intrapleural implant	Stanton et al. [21]

This introduction about tremolite habits and their behavior is directly connected with the purpose of this research: the evaluation of the grinding effect on the morphology of four tremolite samples. This study aims to analyze the morphological changes of both the fibrous and the prismatic states in order to simulate the external process of rock deterioration. The grinding effect strictly depends on both the characteristics of asbestos mineral (rocks cohesion and friability) and interaction of the rock with external "phenomena". In the last year the effect of grinding on tremolite asbestos has been well investigated by Bloise et al. [22] (2018). In this work, the study focuses not only on asbestiform samples but on prismatic tremolite too. This decision arises from the need to verify if the tremolite with prismatic habit, if subjected to a grinding process, can produce potentially harmful fibers. These events can happen during natural degradation processes of rocks (weathering of outcrops) or during anthropic intervention, such as the mechanized excavation or earthmoving process (transport). These processes, which employ a series of crushing and abrasive actions on the rock, are the cause of dimensional reduction of asbestos fibers, where present, and therefore can provoke serious human health problems linked to the release of fibers in air or in groundwater, so entering in the hydrological system [23].

The results showed increases in prismatic elements for all tremolite samples after grinding. Regarding the carcinogenetic studies previously proposed, the increase of prismatic particles could lead to a safer situation in term of reducing impact on both the health of exposed humans (workers, inhabitants, and health professionals) and on the economy and organization for mining companies operating in deposits where a non-asbestos tremolite is present [2].

2. Materials and Methods

Four samples of tremolite were collected in four sites from the north of Italy—three in the Piedmont region (Bracchiello, Monastero di Lanzo, and Caprie) and the last in Aosta Valley (Verrayes). These tremolites showed different initial habits: prismatic for Bracchiello, Caprie, and Verrayes; fibrous for Monastero. For each sample, the particle morphology was investigated before and after grinding, with both a phase contrast optical microscope (PCOM) and a scanning electron microscope (SEM). A counting strategy to differentiate the asbestiform fibers from the non-asbestiform particles was chosen. This approach was focused on a dimensionless parameter, mainly the ratio between length and diameter (L/D), acquired during the PCOM observation, according to the asbestiform definition of the Health and Safety Executive in which an L/D > 20 identifies a particle as fibrous [24].

2.1. Material

The four samples of tremolite analyzed were from Piedmont and Aosta Valley. More precisely, three of four came from Piedmont: Bracchiello (TO), Monastero di Lanzo (TO), and Caprie (TO). The other one, from Aosta Valley, was Verrayes (AO). The samples were collected by Prof. C. Clerici and are part of the mineralogic museum of DIATI.

The sample characteristics are shown in Figure 1.

2.2. Phase Contrast Optical Microscope (PCOM)

The PCOM (Leica Microsystem, Wetzlar, Germany) used was a LEICA Phase Contrast DLMP equipped with 10×, 20× and 40× lenses and a Leica DFC290 digital camera [25]. In order to analyze the material, a small amount of sample was prepared on a microscopic slide and immersed in a high-dispersion liquid with a known refraction index. This refractive index oil affects three parameters: (1) luminosity, (2) color, and (3) birefringence [25–27].

Luminosity is closely connected to the luminance contrast. This means that the brightness ratio between an object and its immediate background depends on an oil with a known refractive index. The color, both of the particle and the surrounding halo, is obtained when the difference between the refractive index of liquids and particles ($\Delta n = n_l - n_p \approx 0$) is within the range −0.020–0.020. The chromatic effect that arises for different Δn and different K (ratio between the slope of the chromatic curves of liquids and solids) is shown in Figure 2, where the color reported above refers to the color of the particle, while the one below refers to the color of the surrounding halo [28].

Sample nomenclature	Aspect	Characteristic
Bracchiello - Bracchiello (TO)	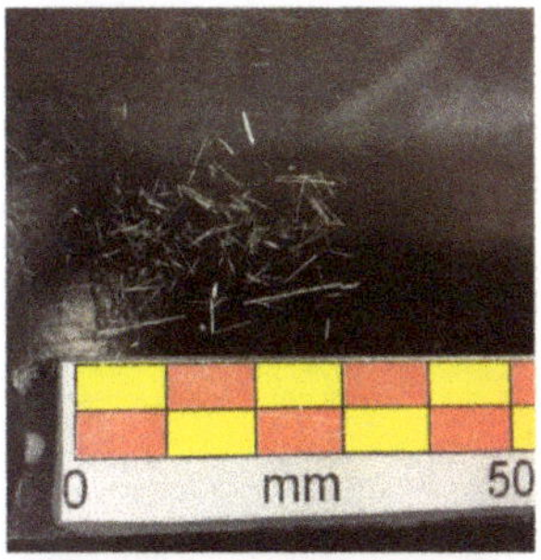	Particles have variable length in the 1–10 mm range and present a rectilinear and rigid aspect (not flexible). They also appear translucent and not aggregate.
Monastero - Monastero di Lanzo (TO)	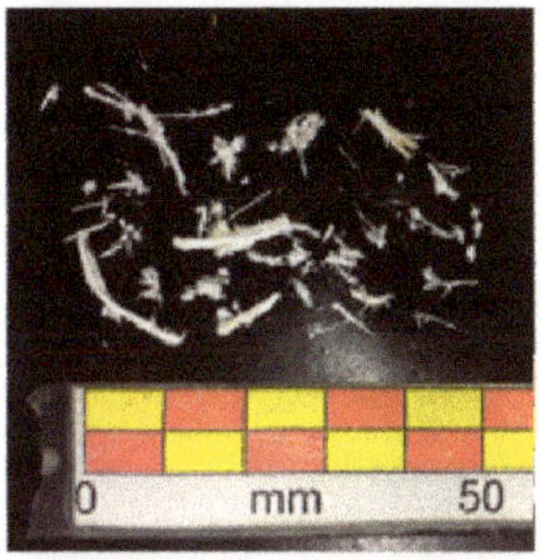	Particles appear aggregated in bundles. This is the only sample that presents a fibrous aspect. Thus, particles have an apparent flexibility and the bundles have frayed extremities.
Caprie - Caprie (TO)	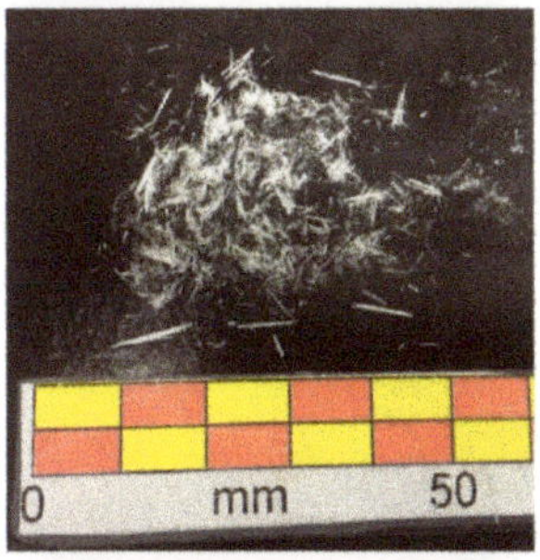	Particles have variable length in the 1–10 mm range. These are prismatic with a rigid and scattered aspect. There is a clear, high quantity of fine components.
Verrayes - Verrayes (AO)	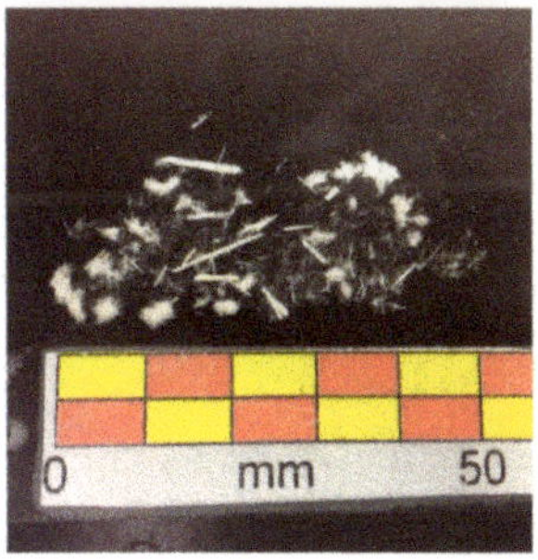	Particles have extreme variability in length and thickness. Prismatic fragments and fine components are noticeable; only the latter are scattered or aggregated in a small mass. There is a clear, high quantity of fine components.

Figure 1. Aspect and characteristics of the samples.

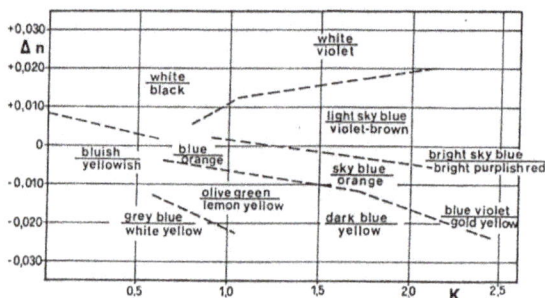

Figure 2. Table of color effects [28] published by Staub, 1955.

2.3. Scanning Electron Microscope (SEM)

The SEM used was a FEI (FEI Company, Hillsboro, OR, USA) operating at 5 and 20 kV. This technology has a resolution of 0.01 μm, for conductive samples, in contrast to the 0.25 μm resolution of the PCOM [25,29]. In this case the resolution of the SEM depends on the grain size of the sputtered coating material (gold) because asbestos is a non-conductive material. It has been used to acquire qualitative images that are subsequently compared with those obtained from the PCOM.

2.4. Methodology

A schematic description of the analysis methodology is reported in Figure 3. For each original sample, two microscopic analyses were carried out: the first on the original samples and the second on the ground samples. The main action of this analysis is to submit both original and ground samples to counting and measuring using the PCOM. SEM was used for a visual comparison to PCOM images.

All the steps will be illustrated in the following subsection.

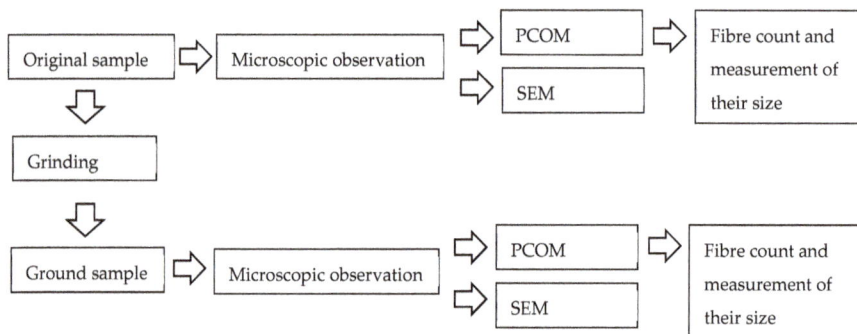

Figure 3. Analysis procedure flowsheet.

2.4.1. Sample Preparations for Microscopic Observation

The first analysis step consisted of the preparation of samples for observation with PCOM and SEM. This action was carried out both for the natural samples and the ground samples.

For the PCOM observation, a portion of the sample was placed on a microscopic slide and immersed in a refractive index oil. To recognize a tremolite asbestos, a liquid with a refraction index equal to 1.615 is suggested [30]. In this study, a liquid with a refraction index of 1.600 was used. This value, slightly lower than the refraction index of the investigated material, allows a maximum contrast between tremolite crystals and the background, reducing the chromatic effect and the halo surrounding the particles. This optical condition is useful for observing the particles with image software programs.

The sample weight must only be considered indicative of a good PCOM observation and particle measurement. In Table 2, the quantity, in milligrams, of material located on the slide is shown.

Table 2. Amount of material on the microscopic slide for each sample.

Sample Nomenclature	Natural Sample (mg)	Ground Sample (mg)
Bracchiello	0.6	0.3
Monastero	0.2	0.3
Caprie	0.3	0.3
Verrayes	0.3	0.3

For SEM observation, the sample must be attached to a stub; each sample was prepared by mixing 10 mg of material in 200 mL of deionized water. Subsequently, 7.5 mL of this mixture was filtered on a polycarbonate membrane (0.4 μm porosity) and fixed on a metallic stub. After that, the specimen must be coated with a thin layer of gold, essentially to abate the increase of high-voltage charges on the specimen and dangerous heat [31].

2.4.2. Grinding

The tremolite behavior after a grinding process was studied by subjecting an amount of original sample, approximately 700 mg, to grinding for 1 minute in an agate jar closed with a sealing gasket lid (model number 952/2 from Humboldt-Wedag). The jar had a hardness of 7.0 Mohs and its dimensions were: diameter of 109 mm, height of 55 mm, and maximum volume of filling of 30 mL.

The fundamental characteristics of the engine's mill were a power of 200 W and a voltage of 230 V/50 Hz. The jar was perfectly fixed and the movement of the crusher was oscillating with eccentric mass.

All the steps of the analysis procedure, such as the selection of the sample, the extraction of the grinding material from the jar, and the preparation of the sample for observation, may lead to a potential exposure to tremolite fibers. Therefore, all steps were carried out using adequate protection devices and under a fume hood (Black Activa Plus from Aquaria).

2.4.3. Decision-Making Processes for Counting and Measurement of Particles

The counting and measurement of each particle were carried out on PCOM images. Eight sample slides were analyzed, four belonging to the original samples and the other four to the grinding samples. For each slide, the number of observation fields had to be selected, as shown in Table 3, and this depended on the object dimensions.

Table 3. Number of fields examined and particles counted for each sample.

Sample Nomenclature	Original Sample		Ground Sample	
	Number of Fields Examined	Number of Particles Counted	Number of Fields Examined	Number of Particles Counted
Bracchiello	100	240	36	331
Monastero	10	2929	25	358
Caprie	100	669	25	288
Verrayes	25	562	25	458

The number of fields investigated differed considerably among the original samples, mainly due to the different dimensions of the particles. A 10× objective has been used for original samples, while a 40× objective for the ground samples. Length and width of particles were measured for each field. The decision-making process is illustrated in Figure 4.

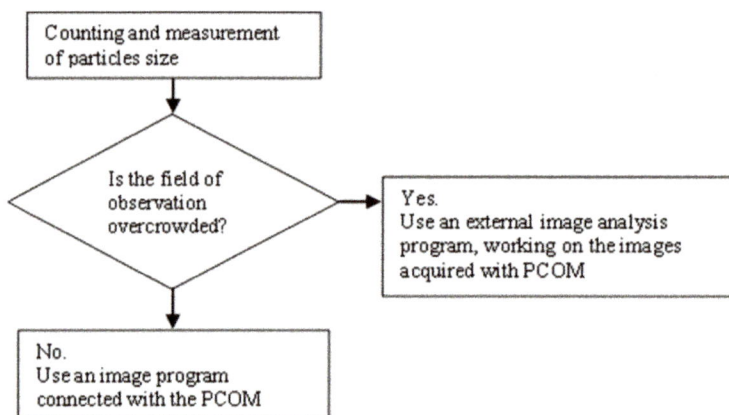

Figure 4. Decision-making process for particle counting and measurement on phase contrast optical microscope (PCOM). Relationship with the field of observation overcrowding.

More precisely, in the case of readable fields of observation (not crowded), real-time measurements were made directly on the image coming from the microscope camera to a monitor. In crowded fields, all the acquired images were instead studied offline with a free image processing software, ImageJ, and both length and thickness were measured. In Figure 5, two images are shown to illustrate field overcrowding, which affected the decision-making process.

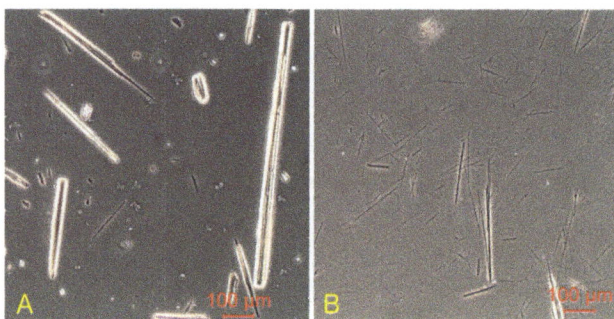

Figure 5. Examples of field crowding for counting and measuring elements: **(A)** Readable field, **(B)** crowded field.

Crystalline particles can be categorized as:

- Fibrous: long and thin fibers;
- Prismatic: elements with a significant thickness and flatness (resulting from planar rupture) or acicular extremities;
- Acicular: long and thin fibers with at least one needle-shaped end;
- Bundle of fibers: indistinguishable elements inside a bundle, where one exists.

The dimensional distinction between fibrous (asbestiform) and prismatic (non-asbestiform) components was made according to the Health and Safety Executive [24]. For HSE "the asbestiform habit is recognized by the following characteristics:

- A range of aspect ratios ranging from 20:1 to 100:1 or higher for fibers longer than 5 μm;
- The capability of splitting into very thin fibrils;

- Two or more of the following:

    - Parallel fibers occurring in bundles;
    - Fiber bundles displaying frayed ends;
    - Fibers in the form of thin needles;
    - Matted masses of individual fibers; and/or
    - Fibers showing curvature."

2.4.4. Closing Remarks on the Counting and Measurement of Particles

The counting and measurement of particles in each field were realized based on the following considerations:

- Only crystalline elements falling within the outline of the microscopic reticle, and not the particles exceeding this area, have been considered, as shown in Figure 6;
- The fields of observation are casually selected inside the coverslip, more precisely following a grid and each field is not repeatable;
- During the measurement, fiber and/or fragment elements having length > 5 μm, without any restriction in the diameter, have been included; 5 μm is a threshold coming from the regulated fibers definition (>5 μm in length, ≤3 μm in diameter width, length/diameter (aspect) ratio ≥ 3:1) according to the World Health Organization and adopted in the Italian Minister Decree 06.09.94 [13,29]. It was chosen to not use the diameter threshold because we needed to measure the width of prismatic components, which are much greater than 3 μm;
- Particles are individually counted but, in the case of a bundle of crystalline elements, where these touch or cross each other it was counted as one fiber;
- In a prismatic fragment, which appears acicular or irregular at one or more point of its length, the diameter is measured along the section. This chosen section must not be influenced by breakage;
- A sufficient number of fields to reach the hundreds of elements were observed.

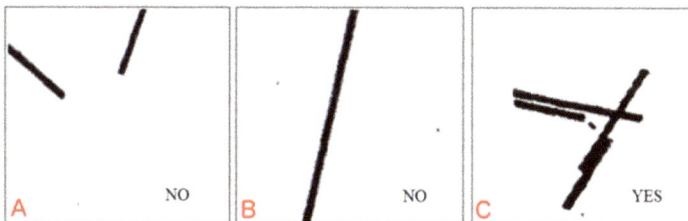

Figure 6. Particle counting: (**A**,**B**) If the elements are cut by the outline of the microscopic reticle, these will be discarded, and (**C**) if the elements fall within the outline of the microscopic reticle, these will be considered [24].

3. Results

Table 4 summarizes all the information about the preparation and observation of samples, ground or not, using PCOM.

Figure 7, PCOM, and Figure 8, SEM, contain the most significant images for both the original and ground samples to give a visual comparison.

Pre-grinding (10x)   Post-grinding (40x)

Bracchiello

Monastero

Caprie

Verrayes

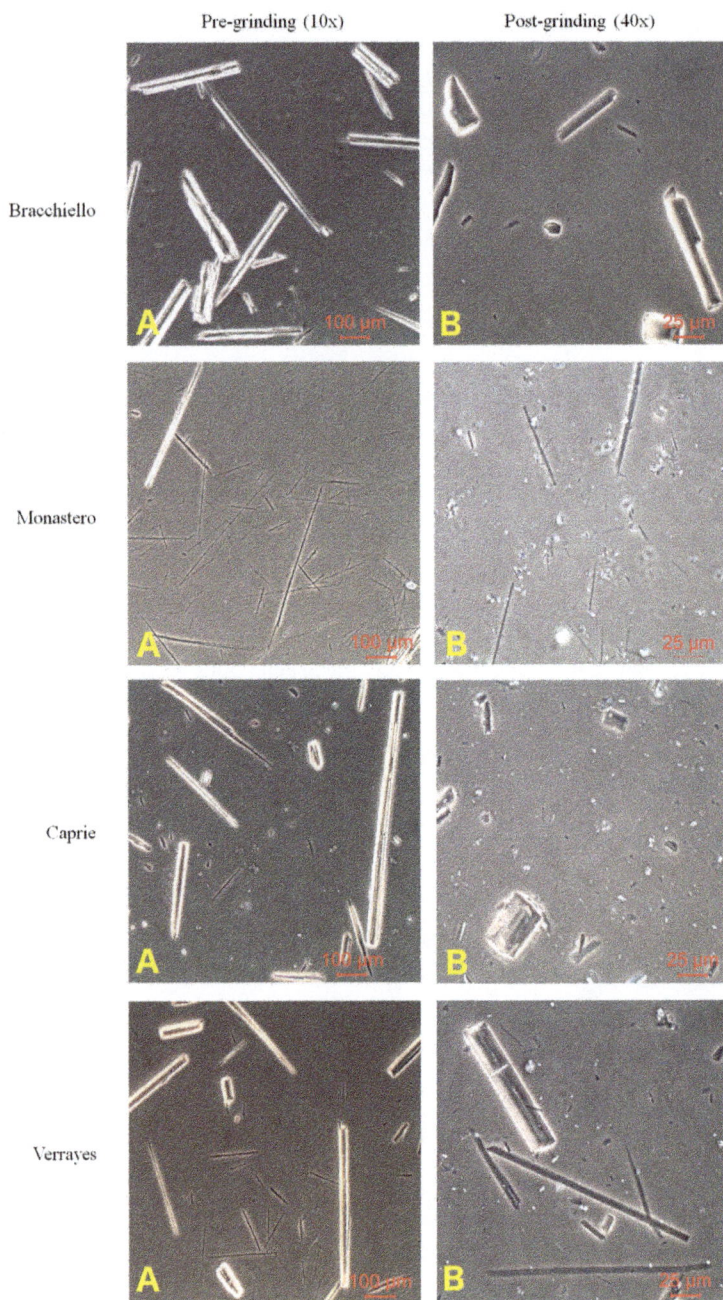

Figure 7. PCOM images (**A**) before and (**B**) after grinding.

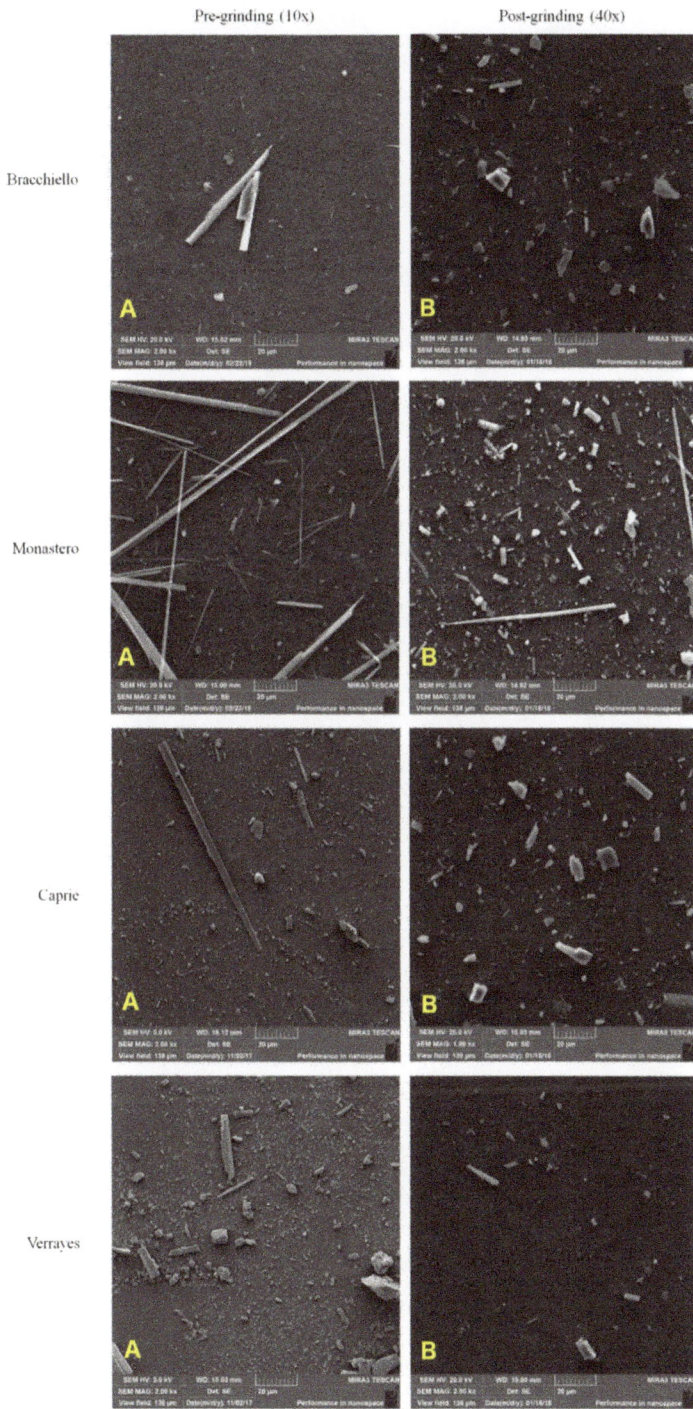

Figure 8. Scanning electron microscope (SEM) images, before (**A**) and after grinding (**B**).

Table 4. Summarized information about original and ground samples.

	Sample nomenclature	Bracchiello	Monastero	Caprie	Verrayes
Original samples	Weight on slide (mg)	0.6	0.2	0.3	0.3
	PCOM objective	10×	10×	10×	10×
	Refractive index of oil	1550	1600	1600	1600
	Fields of observation	100 fields (20 × 5 strips)	20	100 fields (20 × 5 strips)	25
	Coverslip area (mm^2)	25 × 25	34 × 40	25 × 25	25 × 25
	Number of particles analyzed	240	2929	669	526
Ground samples	Sample nomenclature	Bracchiello	Monastero	Caprie	Verrayes
	Weight on slide (mg)	0.3	0.3	0.3	0.3
	PCOM objective	40×	40×	40×	40×
	Refractive index of oil	1600	1600	1600	1600
	Fields of observation	25	25	25	25
	Coverslip area (mm^2)	25 × 25	25 × 25	25 × 25	25 × 25
	Number of particles analyzed	331	358	288	458

The charts in Figure 9 show the results of the granulometric analysis, more precisely the trend in particle length before and after the grinding process. These graphs are frequency histograms, where each class of length is related to its frequency (%). A 5 μm minimum value for length has been chosen according to the minimum value in the respirable fiber definition from WHO [18]. For each class of length, the peak height is defined by the percentage frequency (%) as:

$$\%_i = \frac{(n.of\ fibers)_i}{total\ number\ of\ fibres\ observed} \cdot 100, \tag{1}$$

where i is the ith class of length.

The difference in colors adopted in the graphs shown in Figure 9 helps to recognize the two steps of analysis: yellow for the original samples and green for the ground samples.

The bar charts in Figure 10 illustrate the difference between fibrous and prismatic components before and after grinding based on the Health and Safety Executive definition [29]. A fiber is a component with a length–diameter ratio higher than 20, otherwise, it can be considered a prism.

Therefore, the number of fibrous (L/D > 20) and prismatic (L/D < 20) components were determined for the two-step analysis and their percentage frequency was defined as follows:

$$\%_i fibrous = \frac{\left(n.\ of\ fibers\ with\frac{l}{D} > 20\right)_i}{(total\ number\ of\ fibres\ observed)_i} \cdot 100, \tag{2}$$

$$\%_i prismatic = \frac{\left(n.\ of\ fibers\ with\frac{l}{D} < 20\right)_i}{(total\ number\ of\ fibres\ observed)_i} \cdot 100, \tag{3}$$

where i is the it status of the sample (original or ground).

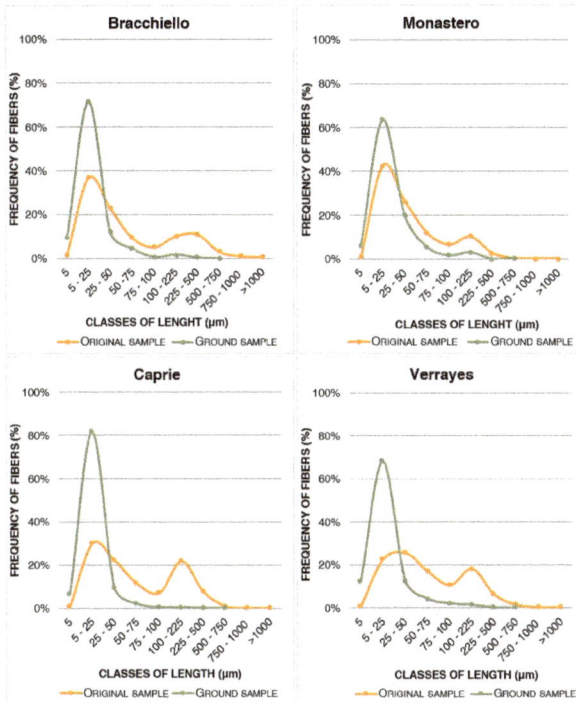

Figure 9. Percentage frequency of particles in each sample, before and after grinding.

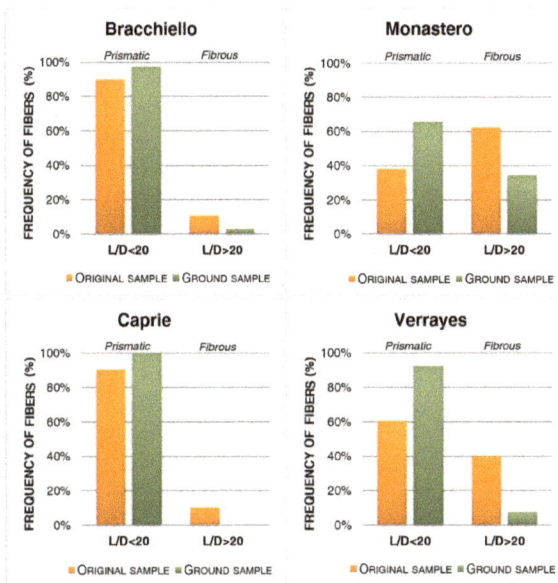

Figure 10. Amounts of fibrous and prismatic component according to the Health and Safety Executive (HSE) [23] before and after grinding.

4. Discussion

4.1. Granulometric Analysis

The results from the granulometric analysis, shown in Figure 9, demonstrate a similarity in the reduction of component length. The original samples are characterized by a great frequency of particles located in the 5–25 μm and 100–225 μm classes. These two classes delineate two obvious peaks and there is one in the 75–100 μm class with a lower frequency. The grinding action is shown by the green curve. This process highlights an attenuation of the second peak (100–225 μm) and a growth in the first (5–25 μm). The long components have a reduction of their length and there is an increase of short elements inside the 5–25 μm class. In short, particles undergo a break perpendicular to the length during grinding and this action causes a reduction of their length. This case is different from the classic asbestos breaking method, which is defined as a longitudinal split of the fibers.

The elements with lengths lower than 5 μm were not considered, because they are not covered in the definition of a respirable fiber by the World Health Organization [13]. The graphs in Figure 6 show a high concentration of ground particles around 25 μm. The maximum diameter is 2.5 μm, ten times higher than the limits of PCOM. Therefore, PCOM is a reliable instrument for this kind of analysis. Moreover, the SEM analysis on the ground samples has permitted checking of the accuracy of the previous analysis. This is especially valuable to avoid errors related to the particle dimensions.

4.2. Dimensional Analysis Based on the Health and Safety Executive Definition

The results of the dimensional analysis, shown in Figure 10, illustrate the variation in the fibrous and prismatic components before and after the grinding process. Looking at the original sample, the yellow bars, Bracchiello and Caprie present more accentuated prismatic components in comparison to the Verrayes sample. The Monastero sample is characterized by the presence of a fibrous component. After the grinding process (the green bars) there is a % decrease of the fibrous elements, with a consequent increase of the prismatic component, which is especially marked for Verrayes and Monastero (decreases of 32% and 27%) and slight for Caprie and Bracchiello (decreases of 9% and 6%).

5. Conclusions

The aim of this study was the evaluation and analysis of tremolite behavior submitted to anthropic (or natural) mechanical actions, which subsequently contribute to the release of fibers into the environment. It is important to underline that tremolite in nature can be present in two crystalline habits: asbestiform (fibrous) or non-asbestiform (prismatic). Therefore, the objective is pointed towards the evaluation of morphology changes in fibers or prism particles after a grinding process. The dimensional analysis, which is based on a direct measurement of the particle size (length and diameter), has allowed the study of these changes by means of the definition from the Health and Safety Executive. HSE distinguishes the fibrous component from the prismatic using the ratio between length and diameter (L/D higher than 20 defines a particle as fibrous).

The results of this research indicate that the prevalent morphology of an amphibolic mineral (in this case, a tremolite asbestos) can change habit. This is the case for the Monastero sample, which initially appeared fibrous but after the grinding process its fibrousness was reduced because the production of prismatic elements took over. The other samples, which initially contained a high content of prismatic particles, were subject to further increases in their prismaticity. The dimensional analysis allowed realistic and reliable results to be obtained for both the samples' composition and for the effects of grinding.

Regarding the carcinogenicity aspects of asbestos, the production of prismatic components in an amphibolic mineral by means of grinding action (digging or earthmoving), can generally be a safer situation for the exposed subjects, such as workers and inhabitants. This is an interesting theme focused on the relationship between the crystalline particle breaking methods and their impact on the respiratory system. In fact, as previously exposed, many experimental studies carried out by different authors proved a lower rate of carcinogenicity for a non-asbestiform amphibole than an asbestiform amphibole.

This study can be used for future surveys, in application to another typology of amphibolic minerals, such as actinolite, and comparing an asbestiform mineral with a non-asbestiform one. Furthermore, it would be also interesting and useful to provide an evaluation about the potential effects on human health and, therefore, obtain a correspondence with the other authors in term of carcinogenicity of these non-asbestiform tremolites.

Author Contributions: Conceptualization: O.B., M.D., P.M. and G.Z.; Data curation: M.D.; Methodology: O.B. and Giovanna Zanetti; Supervision: P.M.; Writing—original draft: M.D.; Writing—review & editing: O.B. and P.M.

Funding: This research received no external funding.

Conflicts of Interest: The authors declare no conflict of interest.

References

1. Deer, W.A.; Howie, R.A.; Zussman, J. *An Introduction to the Rock Forming Minerals*, 1st ed.; Longman Group Limited: London, UK, 1966.
2. Gamble, J.F.; Gibbs, G.W. An evaluation of the risks of lung cancer and mesothelioma from exposure to amphibole cleavage fragments. *Regul. Toxicol. Pharmacol.* **2008**, *52*, S154–S186. [CrossRef] [PubMed]
3. Roggli, V.L.; Vollmer, R.T.; Butnor, K.J.; Sporn, T.A. Tremolite and mesothelioma. *Ann. Occup. Hyg.* **2002**, *45*, 447–453.
4. Addison, J.; McConnell, E.E. A review of carcinogenicity studies of asbestos and non-asbestos tremolite and other amphiboles. *Regul. Toxicol. Pharmacol.* **2008**, *52*, S187–S199. [CrossRef] [PubMed]
5. Ilgren, E.B.; Penna, B.M. The biology of cleavage fragments: A brief synthesis and analysis of current knowledge. *Indoor Built Environ.* **2014**, *13*, 343–356. [CrossRef]
6. Ross, M.; Langer, A.M.; Nord, G.L.; Nolan, R.P.; Lee, R.J.; Orden, D.V.; Addison, J. The mineral nature of asbestos. *Regul. Toxicol. Pharmacol.* **2008**, *52*, S26–S30. [CrossRef] [PubMed]
7. National Research Council. *Asbestiform Fibers: Nonoccupational Health Risks*; National Academies Press: Washington, DC, USA, 1984.
8. Dana, S.D.; Ford, W.E. *A Textbook of Mineralogy*; Wiley and Sons: New York, NY, USA, 1932.
9. Dorling, M.; Zussman, J. Characteristics of asbestiform and non-asbestiform calcic amphiboles. *Lithos* **1987**, *20*, 469–489. [CrossRef]
10. Ministero della Salute Italiana. Sintesi delle conoscenze relative all'esposizione e al profilo tossicologico-Amianto. Available online: http://www.salute.gov.it/imgs/C_17_pubblicazioni_2570_allegato.pdf (accessed on 27 May 2019).
11. IARC. Some inorganic and organometallic compounds. *IARC Monogr. Eval. Carcinog. Risk Chem. Man.* **1973**, *2*, 1–181.
12. Gualtieri, A.F.; Gandolfi, N.B.; Pollastri, S.; Rinaldi, R.; Sala, O.; Martelli, G.; Bacci, T.; Paoli, F.; Viani, A.; Viglituro, R. Assessment of the potential hazard represented by natural raw materials containing minerals fibers—The case of the feldspar from Orani, Sardinia (Italy). *J. Hazard. Mater.* **2018**, *350*, 76–87. [CrossRef]
13. International Programme on Chemical Safety & WHO Task Group on Asbestos and other Natural Mineral Fibers. *Asbestos and Other Natural Mineral Fibers/Published under the Joint Sponsorship of the United Nations Environment Programme*; World Health Organization: Geneva, Switzerland, 1986; Available online: http://www.who.int/iris/handle/10665/37190 (accessed on 27 May 2019).
14. Timbrell, V.; Griffiths, D. Pooley, Possible importance of fiber diameters of South African Amphiboles. *Nature* **1971**, *232*, 55–56. [CrossRef]
15. Wylie, A.; Mossman, B. Mineralogical features associated with cytotoxic and proliferative effects of fibrous talc and asbestos on tracheal epithelial and pleural mesothelial cells. *J. Toxicol. Appl. Pharmacol.* **1997**, *147*, 143–150. [CrossRef]
16. Wagner, J.C.; Chamberlain, M.; Brown, R.; Berry, G.; Pooley, F.; Davies, R.; Griffiths, D. Biological effect of tremolite. *Br. J. Cancer* **1982**, *45*, 352–371. [CrossRef]
17. Davis, J.M.G.; Addison, J.; Bolton, R.E.; Donaldson, K.; Jones, A.D.; Miller, B.G. Inhalation studies on the effects of tremolite and brucite dust in rats. *Carcinogenesis* **1985**, *6*, 667–674. [CrossRef] [PubMed]

18. Wagner, J.C.; Slegs, C.; Marchands, P. Diffuse pleural mesothelioma and asbestos exposure in the North Western Cape Province. *Br. J. Ind. Med.* **1960**, *17*, 260–271. [CrossRef] [PubMed]

19. Smith, W.; Hubert, D.; Sobel, H.; Marquet, E. Biologic tests of tremolite in hamsters. In *Dust and Disease, proceedings of the Conference on Occupational Exposures to Fibrous and Particulate Dust and Their Extension into the Environment, Washington, DC, USA, 1977*; Pathotox Publishers: Park Forest South, IL, USA, 1979; pp. 335–339.

20. Davis, J.M.G.; Addison, J.; McIntosh, C.; Miller, B.; Niven, K. Variations in the carcinogenicity of tremolite dust samples of differing morphology. *Ann. N. Y. Acad. Sci.* **1991**, *643*, 473–483. [CrossRef]

21. Stanton, M.; Layard, M.; Tegeris, A.; Miller, E.; May, M.; Morgan, E.; Smith, A. Relation of particle dimension to carcinogenicity in amphibole asbestos and other fibrous minerals. *J. Natl. Cancel Inst.* **1981**, *67*, 965–975.

22. Bloise, A.; Kusiorowski, R.; Gualtieri, A. The effect of grinding on tremolite asbestos and anthophyllite asbestos. *Minerals* **2018**, *8*, 274. [CrossRef]

23. Gualtieri, A.F.; Pollastri, S.; Gandolfi, N.B.; Ronchetti, F.; Albonico, C.; Cavallo, A.; Zanetti, G.; Marini, P.; Sala, O. Minerals in the human body—Determination of the concentration of asbestos minerals in highly contaminated mine tailings: An example from abandoned mine waste of Crètaz and Èmarese (Valle d'Aosta, Italy). *Am. Mineral.* **2014**, *99*, 1233–1247. [CrossRef]

24. Health and Safety Executive. *Asbestos: The Analysts' Guide for Sampling, Analysis and Clearance Procedures*; HSE Books: Norwich, UK, 2005.

25. Baietto, O.; Marini, P. Naturally occurring asbestos: Validation of PCOM quantitative determination. *Resour. Policy* **2018**, *59*, 44–49. [CrossRef]

26. Clerici, C.; Morandini, A.; Occela, E.; Visetti, A. L'impiego del contrasto di fase in microscopia. *Boll. dell'associ. Miner. Subalp.* **1975**, *12*, 268–298.

27. Niskanen, L.; Räty, J.; Peiponen, K.E. Determination of the refractive index of microparticles by utilizing light dispersion properties of the particle and an immersion liquid. *Talanta* **2013**, *115*, 68–73. [CrossRef] [PubMed]

28. Schmidt, K.G. Die phasenkontrastmikroskopie in der staubtechnik. *Staub* **1955**, *41*, 436.

29. Marconi, A. L'identificazione delle fibre di asbesto per mezzo della tecnica microscopica della dispersione cromatica. *Ann. dell'Ist. Super. Della Sanità* **1982**, *18*, 911–914.

30. Decreto del Ministero della Sanità 6 settembre 1994: Normative e metodologie tecniche di applicazione dell'art. 6, comma 3, e dell'art. 12, comma 2, della legge 27 marzo 1992, n. 257, relativa alla cessazione dell'impiego dell'amianto. (GU Serie Generale n.220 del 20-09-1994 - Suppl. Ordinario n. 129. Available online: https://www.unipd-org.it/rls/pericolirischi/Pericoli/.../d.m._06.09.94.pdf (accessed on 28 May 2019).

31. Bozzola, J.J.; Russell, L.D. *Electron. Microscopy: Principles and Techniques for Biologists*, 2nd ed.; Jones & Bartlett Learning: Burlington, MA, USA, 1992.

MDPI
St. Alban-Anlage 66
4052 Basel
Switzerland
Tel. +41 61 683 77 34
Fax +41 61 302 89 18
www.mdpi.com

Fibers Editorial Office
E-mail: fibers@mdpi.com
www.mdpi.com/journal/fibers

www.ingramcontent.com/pod-product-compliance
Lightning Source LLC
Chambersburg PA
CBHW051914210326
41597CB00033B/6134